FIRST EDITION

A Basic Approach to PreCalculus Trigonometry

Preparing to Succeed in Calculus

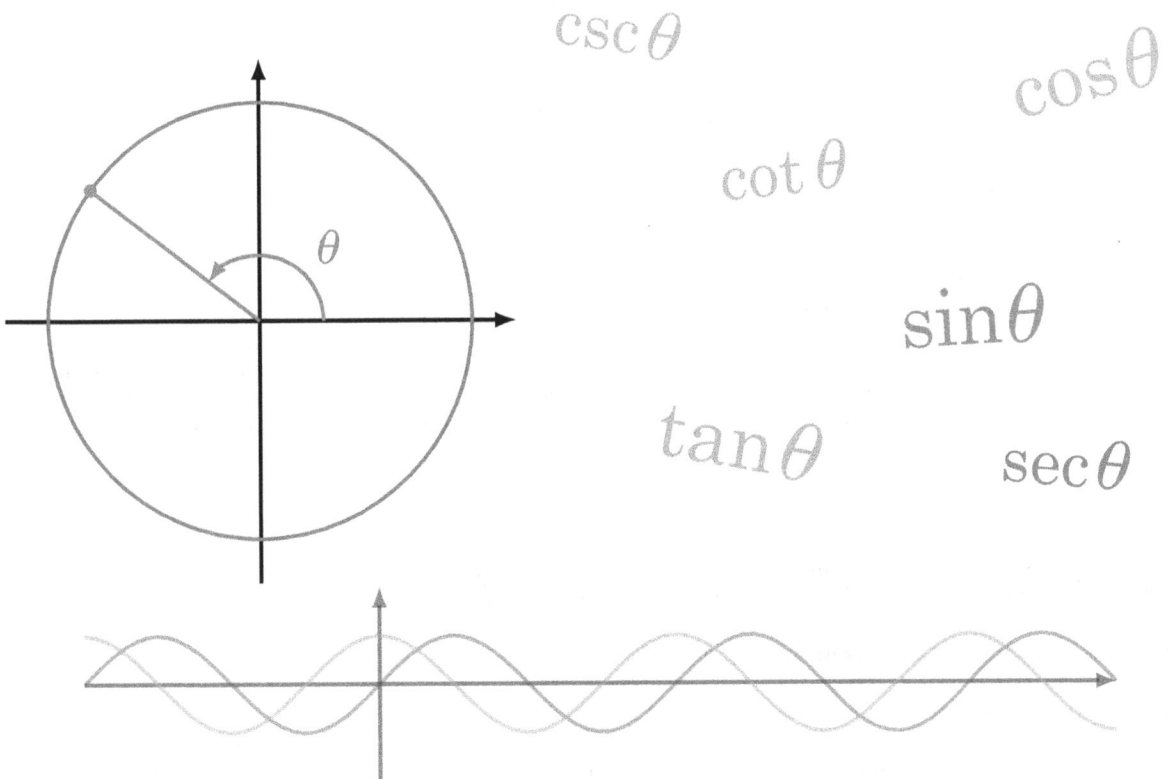

$$\csc\theta \quad \cos\theta \quad \cot\theta \quad \sin\theta \quad \tan\theta \quad \sec\theta$$

Young H. Kim
Youngsoo Kim
Ana M. Tameru
Wen Yan
Tuskegee University

cognella®
academic publishing

Bassim Hamadeh, CEO and Publisher
Michael Simpson, Vice President of Acquisitions and Sales
Jamie Giganti, Senior Managing Editor
Jess Busch, Senior Graphic Designer
Angela Schultz, Senior Field Acquisitions Editor
Michelle Piehl, Project Editor
Alexa Lucido, Licensing Coordinator

First published in the United States of America in 2016 by Cognella, Inc.

Trademark Notice: Product or corporate names may be trademarks or registered trademarks, and are used only for identification and explanation without intent to infringe.

Cover images copyright © Depositphotos/agsandrew.

Printed in the United States of America

ISBN: 978-1-63487-308-6 (pbk) / 978-1-63487-309-3 (br)

cognella®
academic publishing

www.cognella.com   800-200-3908

To our children

Contents

Chapter 1

RATIONAL FUNCTIONS

1.1 Graphing Rational Functions

A **rational function** is a function $f(x)$ that can be written as the ratio of two polynomials $N(x)$ and $D(x)$ where the denominator isn't zero.

$$f(x) = \frac{N(x)}{D(x)}, \quad D(x) \neq 0$$

We use the domain, intercepts, and asymptotes of the function as a guide to draw the graph.

▶ Domain

The **domain** of a function is the set of values of x on which the function is defined. A rational function is simply a fraction, and in a fraction the denominator cannot equal zero because the fraction would be undefined. To find which values of x make the fraction defined, create an equation where the denominator is not equal to zero and solve it. The domain of f is the set of all real numbers x for which $D(x) \neq 0$. The answer is written in interval notation.

Example 1. Finding the Domain of a Rational Function

Find the domain of the rational function.

a) $f(x) = \dfrac{x-1}{x+3}$
b) $g(x) = \dfrac{x-1}{x^2-1}$
c) $h(x) = \dfrac{x-4}{x^2+4}$

Solution.

a) Solve $x + 3 \neq 0$ and get $x \neq -3$. Therefore, the domain of f consists of all real numbers except -3, and it is written as $(-\infty, -3) \cup (-3, \infty)$.

b) Solve $x^2 - 1 \neq 0$ and get $x \neq \pm 1$. Therefore, the domain of f consists of all real numbers except 1 and -1, it is written as $(-\infty, -1) \cup (-1, 1) \cup (1, \infty)$.

c) Solve $x^2 + 4 \neq 0$ or $x^2 \neq -4$. A squared number never equals -4 because squared numbers are always non-negative. Therefore, the domain is "all real numbers," written as $(-\infty, \infty)$.

Practice Now. Find the domain of the rational function.

a) $f(x) = \dfrac{x-2}{x+1}$
b) $g(x) = \dfrac{x+4}{x^2-16}$
c) $h(x) = \dfrac{x+2}{x^2+1}$

▶ Intercepts

The **intercepts** of a function are the intersections of the graph with coordinate axes. The x-intercepts are points of the form $(x, 0)$ on the x-axis, and the y-intercepts are points of the form $(0, y)$ on the y-axis. For rational functions, we find intercepts as follows.

- Find the x-intercept(s) by solving $N(x) = 0$.

- Find the y-intercept by evaluating $f(0)$.

Example 2. Finding Intercepts of a Rational Function

Find the x- and y-intercepts of the rational function.

a) $f(x) = \dfrac{2x - 1}{x + 3}$
b) $g(x) = \dfrac{8}{x^2 + 8}$

Solution.

a) The x-intercepts occur at values for which the numerator is 0. To find the x-intercepts, we solve the equation $2x - 1 = 0$. The solution is $x = \dfrac{1}{2}$. So, the x-intercept is $\left(\dfrac{1}{2}, 0\right)$. To find the y-intercept, we compute $f(0)$.

$$f(0) = \frac{2(0) - 1}{0 + 3} = \frac{-1}{3} = -\frac{1}{3}.$$

So, the y-intercept is $\left(0, -\dfrac{1}{3}\right)$.

b) The x-intercepts occur at values for which the numerator is 0. In this case the numerator is the constant 8, which is never zero. So, there are no x-intercepts. To find the y-intercept, we compute $g(0)$.

$$g(0) = \frac{8}{0^2 + 8} = \frac{8}{8} = 1.$$

The y-intercept is $(0, 1)$.

Practice Now. Find the x- and y-intercepts of the rational function.

a) $f(x) = \dfrac{3x}{8x + 3}$
b) $g(x) = \dfrac{4}{x^2 + 16}$

▶ Asymptotes

An **asymptote** of a curve is a line that the curve comes arbitrarily closer to as it extends farther. There are three kinds of asymptotes—vertical, horizontal, and oblique asymptotes.

Vertical Asymptotes

Assuming that there are no common factors in the numerator and denominator of the function $f(x) = \dfrac{N(x)}{D(x)}$, the vertical asymptotes are obtained by solving the equation $D(x) = 0$. A vertical asymptote is expressed as an equation of the form $x = a$. The vertical line $x = a$ is never touched by the graph.

Example 3. Finding Vertical Asymptotes of a Rational Function Graph

Find the vertical asymptotes, if any.

a) $f(x) = \dfrac{x+5}{x-2}$

b) $g(x) = \dfrac{x+4}{x^2-3x-28}$

c) $f(x) = \dfrac{x-4}{x^2+5}$

Solution.

a) Because there are no common factors in the numerator and denominator, we solve the equation $x - 2 = 0$ to get the only solution, 2. Therefore, the vertical asymptote is $x = 2$.

b) The denominator of the function $g(x)$ is factored, and

$$g(x) = \frac{x+4}{x^2-3x-28} = \frac{\cancel{x+4}}{\cancel{(x+4)}(x-7)} = \frac{1}{x-7}.$$

We solve the equation $x - 7 = 0$ to get the vertical asymptote $x = 7$.

c) Because the denominator $x^2 + 5$ has no real zeros, the graph of $f(x)$ has no vertical asymptote.

Practice Now. Find the vertical asymptotes, if any.

a) $f(x) = \dfrac{3x}{8x+3}$

b) $f(x) = \dfrac{x+1}{x^2+2x+1}$

c) $g(x) = \dfrac{4}{x^2+16}$

Horizontal Asymptotes

The horizontal asymptote is determined using the following rule.

For a rational function

$$f(x) = \frac{ax^n + \cdots}{bx^m + \cdots} \begin{array}{l} \leftarrow n\text{th degree polynomial,} \\ \leftarrow m\text{th degree polynomial,} \end{array} \quad \text{where } a \neq 0,\ b \neq 0.$$

1. If $n < m$, then $y = 0$ (the x-axis) is the horizontal asymptote.

2. If $n = m$, then the horizontal asymptote is the line $y = \dfrac{a}{b}$.

3. If $n > m$, then there is no horizontal asymptote.

Example 4. Finding the Horizontal Asymptote of a Rational Function Graph

Find the horizontal asymptote, if any.

a) $f(x) = \dfrac{x+1}{x^2-2x+3}$

b) $g(x) = \dfrac{7x+4}{28-3x}$

c) $h(x) = \dfrac{x^2-4x+1}{-4x+1}$

Solution.

a) The degree of the numerator is 1, and that of the denominator is 2. Because the degree of the denominator is greater, the horizontal asymptote is $y = 0$ by the first case of the rule.

b) We rearrange terms of the function $g(x)$ in descending order of degrees.

$$g(x) = \frac{7x+4}{28-3x} = \frac{7x+4}{-3x+28}$$

The numerator and the denominator have the same degree of 1. The leading coefficients of the numerator and the denominator are 7 and -3, respectively. By the second case of the rule, the horizontal asymptote is $y = \dfrac{7}{-3} = -\dfrac{7}{3}$.

c) The degree of the numerator is 2, and that of the denominator is 1. Because the degree of the numerator is greater, by the third case of the rule, there is no horizontal asymptote.

Practice Now.   Find the horizontal asymptotes, if any.

a) $f(x) = \dfrac{3x}{8x+3}$ **b)** $f(x) = \dfrac{x+1}{x^2+2x+1}$ **c)** $g(x) = \dfrac{3x^2+16}{x+16}$

Oblique Asymptotes

If the degree of $N(x)$ is exactly one more than the degree of $D(x)$, then the graph of the rational function has an oblique asymptote. Divide the numerator by the denominator using polynomial division to obtain the quotient, which will be of the linear form $mx + b$. The oblique asymptote is $y = mx + b$.

Example 5. Finding the Oblique Asymptote of a Rational Function Graph

Find the oblique asymptote of $f(x) = \dfrac{x^2 - 4x + 5}{x+3}$.

Solution.   Because the degree of the numerator is exactly one more than the degree of the denominator, the graph of $f(x)$ has an oblique asymptote. Performing the long division (or synthetic division) of polynomials, we get the quotient $x - 7$ and remainder 26.

$$f(x) = (x-7) + \frac{26}{x+3}$$

Thus, the oblique asymptote is $y = x - 7$.

Practice Now.   Find the oblique asymptote of $f(x) = \dfrac{2x^2 - 4x + 5}{x+1}$.

▶ Sketching the Graph of a Rational Function

Follow the steps to draw the graph of a rational function $f(x) = \dfrac{N(x)}{D(x)}$.

> ✏ **Sketching the Graph of a Rational Function**
>
> 1. Find the domain.
>
> 2. Find the x- and y-intercepts.
>
> 3. Find vertical and horizontal (or oblique) asymptotes.
>
> 4. Find additional points on the graph of f, particularly near zeros of $N(x)$ and $D(x)$.
>
> 5. Sketch the graph by using the information obtained previously. Draw the asymptotes, plot the intercepts and additional points, and complete the graph with a smooth curve. The number of curves is equal to the number of intervals in the domain.

Example 6. Sketching the Graph of a Rational Function

Sketch the graph of $f(x) = \dfrac{x+2}{x+1}$.

Solution.

Step 1. Find the domain.

Solve $x + 1 \neq 0$ to get $x \neq -1$. The domain of f consists of all real numbers except -1, written in interval notation as $(-\infty, -1) \cup (-1, \infty)$. The domain consists of two intervals, so the graph will consist of two curves, separated by $x = -1$.

Step 2. Find the x- and y-intercepts.

- The x-intercept: Solve $x + 2 = 0$ and get $x = -2$. The x-intercept is $(-2, 0)$.

- The y-intercept: $f(0) = \dfrac{0+2}{0+1} = \dfrac{2}{1} = 2$. So, the y-intercept is $(0, 2)$.

Step 3. Find the asymptotes.

- The vertical asymptote(s): Solve $x + 1 = 0$ and get $x = -1$. Thus, the vertical asymptote is $x = -1$.

- The horizontal asymptote: Because the degrees are the same in the numerator and the denominator, the horizontal asymptote is $y = \dfrac{1}{1} = 1$, obtained by taking the ratio of the leading coefficients.

- The oblique asymptote: The graph of f has no oblique asymptote because both numerator and denominator have the same degrees.

Step 4. Find additional points on the graph of f.

The zeros of numerator and denominator are -2 and -1. Find points near these values.

x	-3	-2	$-\frac{3}{2}$	$-\frac{1}{2}$	0	1
y	$\frac{1}{2}$	0	-1	3	2	$\frac{3}{2}$

Step 5. Graph the function.

- Sketch the vertical asymptote $x = -1$ and the horizontal asymptote $y = 1$ in dotted lines.

- Plot the intercepts $(-2, 0)$ and $(0, 2)$.

- Plot additional points from Step 4.

- Draw two curves by connecting dots and letting them get closer to the asymptotes as they go farther.

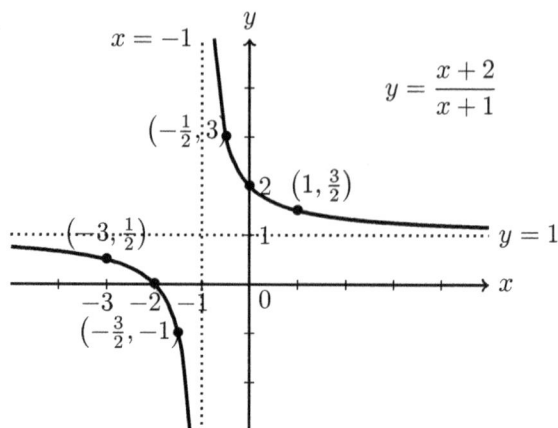

Practice Now.   Sketch the graph of $f(x) = \dfrac{3x^2}{x^2 - 4}$.

Exercises 1.1

(1–6) Find the domain of each rational function.

1. $f(x) = \dfrac{1}{x+2}$

2. $f(x) = \dfrac{x-1}{x+7}$

3. $f(x) = \dfrac{x+1}{(x-2)^2}$

4. $f(x) = \dfrac{x+4}{x^2+x-12}$

5. $f(x) = \dfrac{x^2-4x-5}{x-5}$

6. $f(x) = -\dfrac{4x}{x^2+9}$

(7–12) Find the x- and y-intercepts of the rational function.

7. $f(x) = \dfrac{x-1}{x+2}$

8. $f(x) = \dfrac{3x}{5-x}$

9. $f(x) = \dfrac{x^2+x-2}{x+7}$

10. $f(x) = \dfrac{2}{x^2-3x-4}$

11. $f(x) = \dfrac{5x+4}{x^2+10x+25}$

12. $f(x) = \dfrac{x^2-3x-4}{2x^2+4x-1}$

(13–20) Find the vertical and horizontal asymptotes (if any).

13. $f(x) = \dfrac{x}{x+3}$

14. $f(x) = \dfrac{2x-3}{4-x}$

15. $f(x) = \dfrac{x^2}{x^2-5x-6}$

16. $f(x) = \dfrac{2x+1}{x^2-5x+6}$

17. $f(x) = \dfrac{6x^2+5x+1}{3x^2-5x-2}$

18. $f(x) = \dfrac{x^2-4x-5}{x-12}$

19. $f(x) = \dfrac{x-1}{x^2+8x+15}$

20. $f(x) = \dfrac{5-2x}{(2x+1)(3x+4)}$

(21–26) Find the oblique asymptote, if any, of the graph of each rational function.

21. $f(x) = \dfrac{x^2+4x+4}{x+3}$

22. $f(x) = \dfrac{2x^2+14x+7}{x-5}$

23. $f(x) = \dfrac{x^2}{x^2-6x-6}$

24. $f(x) = \dfrac{2x-1}{x^2+5x+6}$

25. $f(x) = \dfrac{5-2x}{(2x+1)(3x+4)}$

26. $f(x) = \dfrac{2x^6-1}{x^5+3x+1}$

(27–32) Use the method outlined in this section to graph each rational function.

27. $f(x) = \dfrac{x}{x^2-4}$

28. $f(x) = \dfrac{2x^2-7x+6}{x^2-3x-4}$

29. $f(x) = \dfrac{x^2-4}{x-1}$

30. $f(x) = \dfrac{5x+21}{x^2-6x-7}$

31. $f(x) = \dfrac{x+3}{x+1}$

32. $f(x) = \dfrac{2x^3-x^2-x}{x^2-4}$

1.2   Partial Fraction Decomposition

To find the **partial fraction decomposition** of a rational expression $\dfrac{N(x)}{D(x)}$ is to express it as a sum of several fractions, each of which is called a partial fraction, with denominators of smaller degrees.

$$\frac{N(x)}{D(x)} = \frac{N_1(x)}{D_1(x)} + \frac{N_2(x)}{D_2(x)} + \frac{N_2(x)}{D_2(x)} + \cdots + \frac{N_m(x)}{D_m(x)}$$

In this book, we only deal with the "proper" cases where $N(x)$ and $D(x)$ have no common factors and the degree of $N(x)$ is less than the degree of $D(x)$. If this is not the case, we can first perform polynomial division to reduce the degree of $N(x)$. For example, if we were to find the partial fraction decomposition of $\frac{x^3}{x^2 - 1}$, we would perform long division to obtain $\frac{x^3}{x^2 - 1} = x + \frac{x}{x^2 - 1}$. Then we would find the partial fraction decomposition of $\frac{x}{x^2 - 1}$.

The form of partial fractions is determined by the factors of the denominator $D(x)$. Depending on the form of the factors, one of four possible cases will arise.

Case 1) $D(x)$ is a product of distinct linear factors.

Let us assume that $D(x)$ is a product of m distinct linear factors.

$$D(x) = (a_1 x + b_1)(a_2 x + b_2) \cdots (a_m x + b_m).$$

Then the partial fraction decomposition is of the form

$$\frac{N(x)}{D(x)} = \frac{A_1}{a_1 x + b_1} + \frac{A_2}{a_2 x + b_2} + \cdots + \frac{A_m}{a_m x + b_m},$$

where $A_1, ..., A_m$ are constants to be determined.

Remark. Finding the decomposition amounts to solving equations for the constants $A_1, ..., A_m$. This can be done in two different ways. We illustrate them in examples.

Case 2) $D(x)$ has distinct irreducible quadratic factors.

If $D(x)$ contains an irreducible quadratic factor $ax^2 + bx + c$, then the partial fraction for this factor will be of the form

$$\frac{Ax + B}{ax^2 + bx + c},$$

where A, B are constants to be determined.

Remark. A quadratic factor is called irreducible if it cannot be factored into two linear factors. An easy way to check whether it is irreducible is to use the discrminant $b^2 - 4ac$. If $b^2 - 4ac < 0$, then it is irreducible.

Case 3) $D(x)$ has powered linear factors.

Suppose that $D(x)$ contains $(ax + b)^m$, that is, $ax + b$ with the exponent m, where $m \geq 2$. The partial fractions for this factor will be of the form

$$\frac{A_1}{ax + b} + \frac{A_1}{(ax + b)^2} + \cdots + \frac{A_m}{(ax + b)^m},$$

where $A_1, ..., A_m$ are constants to be determined.

Case 4) $D(x)$ has powered irreducible quadratic factors.

Suppose that $D(x)$ contains $(ax^2 + bx + c)^m$, that is, $ax^2 + bx + c$ with the exponent m, where $m \geq 2$. The partial fractions for this factor will be of the form

$$\frac{A_1 x + B_1}{(ax^2 + bx + c)} + \frac{A_2 x + B_2}{(ax^2 + bx + c)^2} + \cdots + \frac{A_m x + B_m}{(ax^2 + bx + c)^m},$$

where A_i and B_i are constants to be determined for $i = 1, 2, \ldots, m$.

✏ Finding the Partial Fraction Decomposition

1. Factor the denominator and set up the form of partial fraction decomposition with unknown constants $A, B, C, \ldots$ in the numerator of the decomposition.

2. Multiply both sides of the resulting equation by the denominator of the left-hand side (the original denominator).

3. Remove the parenthesis using the distributive property. Group like terms on the right-hand side.

4. Write both sides in descending powers, equate coefficients of like powers of x, and equate constant terms.

5. Solve the system of linear equations for $A, B, C, \ldots$

6. Substitute the values $A, B, C, \ldots$ into the equation in Step 1 and write the partial fraction decomposition.

Example 1. Find the Partial Fraction Decomposition with Linear Factors

Find the partial fraction decomposition of $\dfrac{2x + 1}{x^2 + 3x + 2}$.

Solution.

- Step 1: We begin by factoring the denominator. We obtain $\dfrac{2x + 1}{(x + 1)(x + 2)}$. The denominator is the product of linear factors. So, we have

$$\frac{2x + 1}{(x + 1)(x + 2)} = \frac{A}{(x + 1)} + \frac{B}{(x + 2)}.$$

We need to determine the values of A and B.

- Step 2: Multiply both sides by the denominator $(x + 1)(x + 2)$. We obtain

$$\cancel{(x + 1)(x + 2)} \times \frac{2x + 1}{\cancel{(x + 1)(x + 2)}} = \left[\frac{A}{(x + 1)} + \frac{B}{(x + 2)} \right] \times (x + 1)(x + 2).$$

- Step 3: Simplify the right-hand side of the equation using the distributive property.

$$2x + 1 = A(x + 2) + B(x + 1)$$

- Step 4: Expand and write both sides in descending power of x, then combine like terms.

$$2x + 1 = Ax + 2A + Bx + B$$
$$2x + 1 = (Ax + Bx) + (2A + B) \qquad \text{(Rearrange terms.)}$$
$$2x + 1 = (A + B)x + (2A + B) \qquad \text{(Combine like terms.)}$$

Two polynomials are equal if their corresponding coefficients are equal. This gives us the following system.

$$\begin{cases} A + B = 2 \\ 2A + B = 1 \end{cases}$$

- Step 5: Solve the equations for A and B. You may use any method such as the substitution method, the elimination method, or Cramer's rule. The solution of this system is $A = -1$ and $B = 3$.

- Step 6: Substitute the values of A and B into the form in Step 1 and write the partial fraction decomposition to get the answer.

$$\frac{A}{(x+1)} + \frac{B}{(x+2)} = \frac{-1}{(x+1)} + \frac{3}{(x+2)}$$
$$= -\frac{1}{(x+1)} + \frac{3}{(x+2)}$$

Remark. There is an alternative way to find A and B by *substituting well-chosen values for x* in this case. The solution is the same until Step 3. We begin with the setup.

$$\frac{2x+1}{(x+1)(x+2)} = \frac{A}{(x+1)} + \frac{B}{(x+2)}$$

We multiply both sides by the denominator of the left-hand side, and simplify to obtain the equation in Step 3.

$$2x + 1 = A(x+2) + B(x+1).$$

From here, we select values for x, which will make all but one of the coefficients go away. We will then be able to solve for that coefficient. More precisely,

- When $x = -1$, we obtain

$$2(-1) + 1 = A(-1+2) + B(-1+1)$$
$$-1 = A \qquad \text{(Simplify.)}$$
$$A = -1 \qquad \text{(Get the solution.)}$$

- When $x = -2$, we obtain:

$$2(-2) + 1 = A(-2+2) + B(-2+1)$$
$$-3 = -B \qquad \text{(Simplify.)}$$
$$B = 3 \qquad \text{(Get the solution.)}$$

Therefore, the decomposition is

$$\frac{A}{(x+1)} + \frac{B}{(x+2)} = \frac{-1}{(x+1)} + \frac{3}{(x+2)}$$

$$= -\frac{1}{(x+1)} + \frac{3}{(x+2)}$$

Practice Now.   Find the partial fraction decomposition of $\dfrac{x+14}{x^2-6x+8}$.

Example 2. Finding the Partial Fraction Decomposition with Irreducible Quadratic Factors

Find the partial fraction decomposition of $\dfrac{3x^2+2x-6}{x(x^2+2x+6)}$.

Solution.

- Step 1: The denominator is already in factored form. Note that the quadratic factor is irreducible because $b^2 - 4ac = 2^2 - 4(1)(6) = -20$ is negative. Set up the form of the partial fraction decomposition.

$$\frac{3x^2+2x-6}{x(x^2+2x+6)} = \frac{A}{x} + \frac{Bx+C}{x^2+2x+6}$$

- Step 2: Multiply both sides by the denominator $x(x^2+2x-6)$ of the left-hand side.

$$\cancel{x(x^2+2x+6)} \times \frac{3x^2+2x-6}{\cancel{x(x^2+2x+6)}} = \left(\frac{A}{x} + \frac{Bx+C}{x^2+2x+6}\right) \times x(x^2+2x+6)$$

- Step 3: Simplify the right-hand side using the distributive property.

$$3x^2+2x-6 = A(x^2+2x+6) + (Bx+C)x$$

- Step 4: Expand and group like terms on the right-hand side.

$$
\begin{aligned}
3x^2+2x-6 &= Ax^2 + 2Ax + 6A + Bx^2 + Cx \\
3x^2+2x-6 &= (Ax^2 + Bx^2) + (2Ax + Cx) + 6A && \text{(Rearrange terms.)}\\
3x^2+2x-6 &= (A+B)x^2 + (2A+C)x + 6A && \text{(Combine like terms.)}
\end{aligned}
$$

  Then equate coefficients.

$$
\begin{cases}
A + B & = 3 \\
2A + C & = 2 \\
6A & = -6
\end{cases}
$$

- Step 5: Solve for $A, B,$ and C. The solution of this system is $A = -1$, $B = 4$, and $C = 4$.

- Step 6: By substituting the values into the form in Step 1, we have the partial fraction decomposition.

$$
\begin{aligned}
\frac{A}{x} + \frac{Bx+C}{x^2+2x+6} &= \frac{-1}{x} + \frac{4x+4}{x^2+2x+6} \\
&= -\frac{1}{x} + \frac{4x+4}{x^2+2x+6}
\end{aligned}
$$

Practice Now.   Find the partial fraction decomposition of $\dfrac{x^2+5x+4}{(x-2)(x^2+2)}$.

Example 3. Finding the Partial Fraction Decomposition with Powered Linear Factors

Find the partial fraction decomposition of $\dfrac{-x^2 + 3x + 1}{x(x+1)^2}$.

Solution.

- Step 1: The denominator is already factored. Note that $x + 1$ is squared. The powered factor will introduce multiple partial fractions with increasing exponents in the denominator. Set up the form of the partial fraction decomposition.

$$\frac{-x^2 + 3x + 1}{x(x+1)^2} = \frac{A}{x} + \frac{B}{x+1} + \frac{C}{(x+1)^2}$$

- Step 2: Multiply both sides by the denominator of the left-hand side $x(x+1)^2$.

$$\cancel{x(x+1)^2} \times \frac{-x^2 + 3x + 1}{\cancel{x(x+1)^2}} = \left[\frac{A}{x} + \frac{B}{x+1} + \frac{C}{(x+1)^2}\right] \times x(x+1)^2$$

- Step 3: Simplify the right-hand side of the equation using the distributive property.

$$-x^2 + 3x + 1 = A(x+1)^2 + Bx(x+1) + Cx$$

We need to find the values of A, B, and C.

- Step 4: We select values for x, which will make all but one of the coefficients go away.

  - When $x = 0$, we obtain:

$$-(0)^2 + 3(0) + 1 = A(0+1)^2 + B(0)(0+1) + C(0)$$
$$1 = A \qquad\qquad \text{(Simplify.)}$$
$$A = 1 \qquad\qquad \text{(Solve.)}$$

  - When $x = -1$, we obtain:

$$-(-1)^2 + 3(-1) + 1 = A(-1+1)^2 + B(-1)(-1+1) + C(-1)$$
$$-3 = -C \qquad\qquad \text{(Simplify.)}$$
$$C = 3 \qquad\qquad \text{(Solve.)}$$

  - To find B, we choose for x any small number, say, 1, then use already found values of A and C.

$$-1^2 + 3(1) + 1 = A(1+1)^2 + B(1)(1+1) + C(1)$$
$$3 = 4A + 2B + C \qquad\qquad \text{(Simplify.)}$$
$$3 = 4(1) + 2B + 3 \qquad\qquad \text{(Substitute.)}$$
$$B = -2 \qquad\qquad \text{(Solve.)}$$

- Step 5: Substitute the values $A = 1$, $B = -2$, and $C = 3$ into the form in Step 1 and write the partial fraction decomposition.

$$\frac{A}{x} + \frac{B}{x+1} + \frac{C}{(x+1)^2} = \frac{1}{x} + \frac{-2}{x+1} + \frac{3}{(x+1)^2}$$
$$= \frac{1}{x} - \frac{2}{x+1} + \frac{3}{(x+1)^2}$$

Practice Now.   Find the partial fraction decomposition of $\dfrac{5x^2 - 8x + 2}{x^3 - 2x^2 + x}$.

Example 4. Finding the Partial Fraction Decomposition with Powered Irreducible Quadratic Factors

Find the partial fraction decomposition of $\dfrac{x^2}{(x^2 + 9)^2}$.

Solution.

- Step 1: The denominator is already factored. Set up the form of the partial fraction decomposition.

$$\frac{x^2}{(x^2 + 9)^2} = \frac{Ax + B}{x^2 + 9} + \frac{Cx + D}{(x^2 + 9)^2}$$

- Step 2: Multiply both sides by the denominator of the left-hand side $(x^2 + 9)^2$.

$$\cancel{(x^2 + 9)^2} \times \frac{x^2}{\cancel{(x^2 + 9)^2}} = \left(\frac{Ax + B}{x^2 + 9} + \frac{Cx + D}{(x^2 + 9)^2} \right) \times (x^2 + 9)^2$$

- Step 3: Simplify the right-hand side of the equation using the distributive property.

$$x^2 = (Ax + B)(x^2 + 9) + (Cx + D)$$

- Step 4: Expand and group like terms on the right-hand side.

$$
\begin{aligned}
x^2 &= Ax^3 + 9Ax + Bx^2 + 9B + Cx + D & \text{(Expand.)} \\
x^2 &= Ax^3 + Bx^2 + (9Ax + Cx) + (9B + D) & \text{(Rearrange terms.)} \\
x^2 &= Ax^3 + Bx^2 + (9A + C)x + (9B + D) & \text{(Combine like terms.)}
\end{aligned}
$$

Then equate coefficients.

$$\begin{cases} A = 0 \\ B = 1 \\ 9A + C = 0 \\ 9B + D = 0 \end{cases}$$

- Step 5: Solve the system of linear equations for $A, B, C,$ and D.
  We get $A = 0$, $B = 1$, $C = 0$, and $D = -9$.

- Step 6: Substitute the values into the form in Step 1 and write the partial fraction decomposition.

$$
\begin{aligned}
\frac{Ax + B}{x^2 + 9} + \frac{Cx + D}{(x^2 + 9)^2} &= \frac{0 \cdot x + 1}{x^2 + 9} + \frac{0 \cdot x + (-9)}{(x^2 + 9)^2} \\
&= \frac{1}{x^2 + 9} - \frac{9}{(x^2 + 9)^2}
\end{aligned}
$$

Practice Now.   Find the partial fraction decomposition of $\dfrac{x^2 + 5x + 4}{(x - 2)(x^2 + 2)}$.

Exercises 1.2

(1–8) Write the form of the partial fraction decomposition of the function. Do not determine the constants.

1. $\dfrac{1}{(x-1)(x+2)}$

2. $\dfrac{x}{x^2+3x-4}$

3. $\dfrac{3x+5}{x(x+2)^2}$

4. $\dfrac{1}{(x-3)(x^2+4)}$

5. $\dfrac{9x-2}{(x^2+1)(x^2+2)}$

6. $\dfrac{7x+3}{x^3+2x^2-3x}$

7. $\dfrac{4x-3}{x^2(x^2+4)^2}$

8. $\dfrac{4x-3}{x^3(x^2+x+1)^3}$

(9–16) Find the partial fraction decomposition of the rational function.

9. $\dfrac{1}{x(x+1)}$

10. $\dfrac{15x-42}{x^2-6x+8}$

11. $\dfrac{2}{9x^2-16}$

12. $\dfrac{3x-28}{(x-7)^2}$

13. $\dfrac{2x^2-23x+57}{(x+3)(x-3)^2}$

14. $\dfrac{12x^2-15x+6}{(x-2)(x^2+2)}$

15. $\dfrac{6x^2}{(x^2+2)^2}$

16. $\dfrac{3x^4+x^3+20x^2+3x+31}{(x+1)(x^2+4)^2}$

Chapter 2

TRIGONOMETRIC FUNCTIONS

2.1 Angles and Their Measurement

▶ Angles in Standard Position

An **angle** is the figure formed by two rays or line segments that have a common endpoint. The angle is measured from the **initial side** to the **terminal side**. The common endpoint is called the **vertex** of the angle.

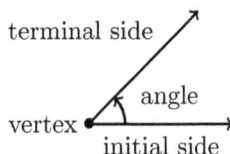

An angle is in **standard position** if it is drawn on the xy-plane with its vertex at the origin and its initial side along the positive x-axis. A **positive angle** rotates counterclockwise, and a **negative angle** rotates clockwise.

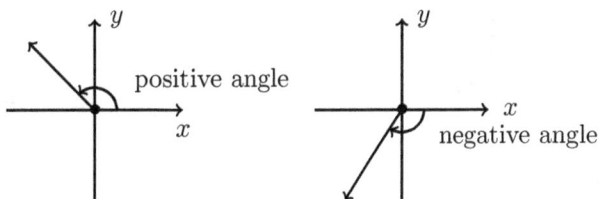

▶ Measurement of Angles

Two ways to measure the size of an angle are with degrees and radians.

Degree Measure

An angle formed by one complete counterclockwise rotation has measure 360 degrees, denoted 360°. Therefore, $1° = \dfrac{1}{360}$ of a complete revolution

---

Example 1. Drawing an Angle in Standard Position

Draw a brief sketch of each angle in standard position.

  a) 150° **b)** −120°

Solution.

  a) A 150° angle is $\dfrac{150°}{360°} = \dfrac{5}{12}$ rotation counterclockwise.

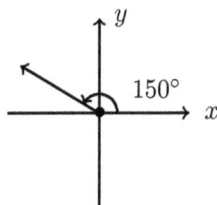

b) A $-120°$ angle is $\dfrac{120°}{360°} = \dfrac{1}{3}$ rotation clockwise.

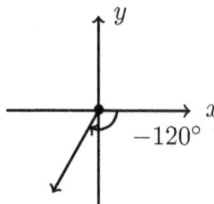

Practice Now.   Draw a brief sketch of the following angles in standard position.

a) $330°$ **b)** $-240°$

Each degree is divided into minutes and seconds. The degree is divided into 60 minutes ($'$). For even finer measurements the minute is divided again into 60 seconds ($''$).

> ✏️ **Relationships between Degrees, Minutes, and Seconds**
>
> We use the following rules for conversion between units.
>
> $$1° = 60' \qquad\qquad 1' = 60'' \qquad\qquad 1° = 3600''$$
>
> $$1' = \left(\frac{1}{60}\right)° \qquad\qquad 1'' = \left(\frac{1}{60}\right)' \qquad\qquad 1'' = \left(\frac{1}{3600}\right)°$$

Example 2. Converting Angles from DMS to DD

Write $46°36'18''$ in DD (decimal degrees) notation.

Solution.

1. The degrees are in whole number: $46°$.

2. Divide the minutes by 60: $36' = \left(\dfrac{36}{60}\right)° = 0.6°$.

3. Divide the seconds by 3600: $18'' = \left(\dfrac{18}{3600}\right)° = 0.005°$.

4. Add all: $46° + 0.6° + 0.005° = 46.605°$.

5. Answer: $46°36'18'' = 46.605°$

Practice Now.   Convert the angle to DD notation.

a) $50°18'40''$ **b)** $26°23'50''$

Example 3. Converting Angles from DD to DMS

Write $38.347°$ in DMS (in degrees, minutes, and seconds) notation.

Solution.

1. The integer part of the number is the degrees: $38°$.

2. Multiply the decimal part by 60: $0.347 \times 60 = 20.82$.

3. The integer part of the number is the minutes: $20'$.

4. Multiply the decimal part by 60: $.82 \times 60 = 49.2$. Round to the nearest whole number, in this case 49. This represents the seconds: $49''$.

5. Answer: $38.347° = 38°20'49''$

Practice Now.   Convert the angle to DMS notation.

    a) $67.634°$ **b)** $29.032°$

Radians

The **radian** measure of an angle θ, which intercepts an arc in a circle, is equal to the ratio of the arc length to the radius of the circle; that is, $\theta = \dfrac{s}{r}$, where s is the arc length, and r is the radius.

> ✏️ **Relationships between Degrees and Radians**
>
> $$180° = \pi \text{ rad}, \qquad 1 \text{ rad} = \left(\frac{180}{\pi}\right)°, \qquad 1° = \frac{\pi}{180} \text{ rad}.$$
>
> - To convert degrees to radians, multiply degrees by $\dfrac{\pi \text{ rad}}{180°}$.
>
> - To convert radians to degrees, multiply radians by $\dfrac{180°}{\pi \text{ rad}}$.

Example 4. Conversion from Degrees to Radians

Convert each angle in degrees to radians.

    a) $45°$ **b)** $-120°$

Solution.   Multiply the angle by $\dfrac{\pi}{180°}$ rad, and simplify the answer.

a) $45° = 45° \cdot \dfrac{\pi}{180°} \text{ rad} = \cancel{45°} \cdot \dfrac{\pi}{4(\cancel{45°})} \text{ rad} = \dfrac{\pi}{4} \text{ rad}$

b) $-120° = -120° \cdot \dfrac{\pi}{180°} \text{ rad} = -2(\cancel{60°}) \cdot \dfrac{\pi}{3(\cancel{60°})} \text{ rad} = -\dfrac{2\pi}{3} \text{ rad}$

Practice Now.   Convert each angle in degrees to radians.

    a) $135°$ **b)** $-225°$

Example 5. Conversion from Radians to Degrees

Convert each angle in radians to degrees.

a) $\dfrac{4\pi}{3}$ **b)** $-\dfrac{7\pi}{15}$

Solution.   Multiply the angle by $\dfrac{180°}{\pi \text{ rad}}$ and simplify the answer.

a) $\dfrac{4\pi}{3}$ rad $= \left(\dfrac{4\pi}{3}\text{ rad}\right)\left(\dfrac{180°}{\pi \text{ rad}}\right) = 240°$ **b)** $-\dfrac{7\pi}{15}$ rad $= \left(-\dfrac{7\pi}{15}\text{ rad}\right)\left(\dfrac{180°}{\pi \text{ rad}}\right) = -84°$

Practice Now.   Convert each angle in radians to degrees.

a) $\dfrac{3\pi}{8}$ **b)** $-\dfrac{8\pi}{9}$

▶ **Types of Angles**

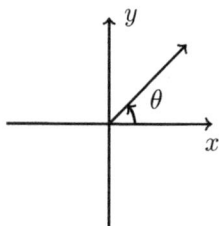   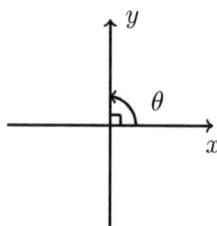   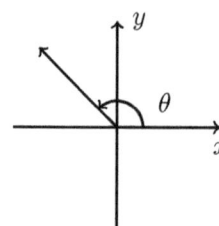

| **acute** angle | **right** angle | **obtuse** angle |
| $(0° < \theta < 90°)$ | $(\theta = 90°)$ | $(90° < \theta < 180°)$ |

▶ **Complements and Supplements**

Two positive angles are **complementary** if their sum equals 90°. They are **supplementary** if the sum equals 180°. If one angle is known, its complementary angle (complement) and supplementary angle (supplement) can be found by subtracting the known angle from 90° and 180°, respectively.

Example 6. Finding the Complement and Supplement of an Angle

Find the complement and the supplement for angle 78°.

Solution.   The complement of 78° is 90° − 78° = 12°. The supplement is 180° − 78° = 102°.

Practice Now.   Find the complement and the supplement for angle 47°.

▶ The Length of a Circular Arc

An arc of a circle is a "portion" of the circumference of the circle.
The length of an arc of a circle is proportional to its central angle.

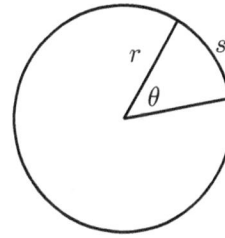

✏️ **The Length of a Circlular Arc**

In a circle of radius r, the length s of an arc that subtends a central angle of θ radians is

$$s = r\theta$$

Example 7. Finding the Length of an Arc

Find the length of an arc in a circle of radius 10 meters that subtends a central angle of $30°$.

Solution.   Convert the angle to radians: $30° = \dfrac{\pi}{6}$ rad. Then the length of the arc is

$$s = r\theta = 10\left(\frac{\pi}{6}\right) = \frac{5\pi}{3} \approx 5.24 \text{ meters}$$

Practice Now.   In a circle of radius 10 centimeters, an arc is intercepted by a central angle with measure $135°$. Find the arc length.

✏️ **The Area of a Sector**

In a circle of radius r, the area A of a sector with a central angle of θ radians is

$$A = \frac{1}{2}r^2\theta.$$

Example 8. Finding the Area of a Sector

Find the area of a sector of a circle with central angle $45°$ if the radius of the circle is 3 meters.

Solution.   Convert the angle to radians: $45° = 45°\left(\dfrac{\pi}{180°} \text{ rad}\right) = \dfrac{\pi}{4}$ rad. Then the area of the sector is

$$A = \frac{1}{2}r^2\theta = \frac{1}{2}(3)^2\left(\frac{\pi}{4}\right) = \frac{9\pi}{8} \approx 3.53 \text{ square meters}$$

Practice Now.   Find the area of the sector associated with a single slice of pizza when the pizza, with a diameter of 14 inches, is cut into eight equal pieces.

▶ Linear and Angular Speed

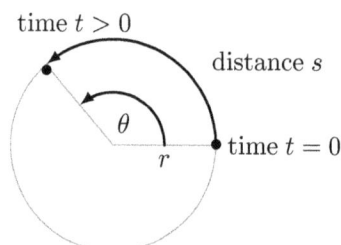

time $t > 0$

distance s

θ

r

time $t = 0$

Suppose that an object moves along a circle of radius r, traveling a distance s over a period of time t, as in the figure. Let θ be the angle swept by the object in that period of time. Then we define the (average) **linear speed** v and the (average) **angular speed** w of the object as:

✎ Linear Speed and Angular Speed

- The (average) linear speed: $v = \dfrac{s}{t}$

- The (average) angular speed: $w = \dfrac{\theta}{t}$

- $v = \dfrac{s}{t} = \dfrac{r\theta}{t} = \dfrac{\theta}{t} \cdot r = wr$

Example 9. Finding Linear Speed and Angular Speed

An object sweeps out a central angle of $\dfrac{\pi}{3}$ radians in 0.5 seconds as it moves along a circle of radius 3 m. Find its linear and angular speed over that time period.

Solution.   We have $t = 0.5$ sec, $r = 3$ m, and $\theta = \dfrac{\pi}{3}$ rad. So, the angular speed is

$$w = \frac{\theta}{t} = \frac{\frac{\pi}{3} \text{ rad}}{0.5 \text{ sec}} = \frac{2\pi}{3} \text{ rad/sec},$$

and the linear speed is

$$v = wr = \left(\frac{2\pi}{3} \text{ rad/sec} \right) (3 \text{ m}) = 2\pi \text{ m/sec}.$$

Practice Now.   An object travels a distance of 35 feet in 2.7 seconds as it moves along a circle of radius 2 feet. Find its linear and angular speed over that time period.

Exercises 2.1

(1–8) Draw a brief sketch of the following angles in standard position. (Do not use a protractor.)

1. $60°$

2. $225°$

3. $-180°$

4. $-300°$

5. $\dfrac{4\pi}{3}$

6. $\dfrac{11\pi}{3}$

7. $-\dfrac{5\pi}{6}$

8. $-\dfrac{4\pi}{3}$

(9–10) Write each angle in DD notation. Round your answers to the nearest hundredth.

9. $50°18'45''$

10. $78°25'8''$

(11–12) Write each angle in DMS notation. Round your answers to the nearest hundredth.

11. $45.5125°$

12. $50.5228°$

(13–21) Convert from degrees to radians. Leave the answers in terms of π.

13. $20°$

14. $155°$

15. $225°$

16. $270°$

17. $-60°$

18. $-120°$

19. $-240°$

20. $-315°$

21. $-330°$

(22–29) Convert from radians to degrees.

22. $\dfrac{\pi}{6}$

23. $\dfrac{7\pi}{6}$

24. $\dfrac{3\pi}{8}$

25. $\dfrac{3\pi}{4}$

26. $-\dfrac{5\pi}{12}$

27. -6π

28. $-\dfrac{2\pi}{9}$

29. $-\dfrac{\pi}{15}$

(30–35) Find, if possible, the complement and the supplement of the given angle.

30. $19°$

31. $42°$

32. $89°$

33. $75°$

34. $170°$

35. $275°$

(36–39) Find the exact length of the arc made by the indicated central angle and radius of each circle.

36. $\theta = \dfrac{\pi}{12}$, $r = 8$ ft

37. $\theta = \dfrac{3\pi}{4}$, $r = 4$ yd

38. $\theta = 22°$, $r = 20$ m

39. $\theta = 210°$, $r = 5$ cm

(40–43) Find the area of the circular sector given the indicated radius and central angle. Round your answers to the neartest tenth.

40. $\theta = \dfrac{\pi}{6}$, $r = 7$ ft

41. $\theta = \dfrac{3\pi}{8}$, $r = 1.2$ yd

42. $\theta = 56°$, $r = 2.1$ m

43. $\theta = 210°$, $r = 6$ cm

(44–45) Find the linear speed of a point traveling at a constant speed along a circle with radius r and angular speed w.

44. $w = \dfrac{2\pi}{3}$ rad/sec, $r = 9$ in

45. $w = \dfrac{\pi}{20}$ rad/sec, $r = 5$ in

(46–47) Find the angular speed of a point traveling at a constant speed along a circle with radius r and linear speed v.

46. $v = 3$ ft/sec, $r = 9$ in

47. $v = 14$ in/sec, $r = 6$ in

2.2 Right Triangle Trigonometry

▶ **The Pythagorean Theorem and Trigonometric Functions**

The Pythagorean theorem states that in any right triangle, the square of the hypotenuse is equal to the squared sum of the lengths of two other sides. It implies that given the lenghts of any two sides of a right triangle, we can find the third as in the following formulas.

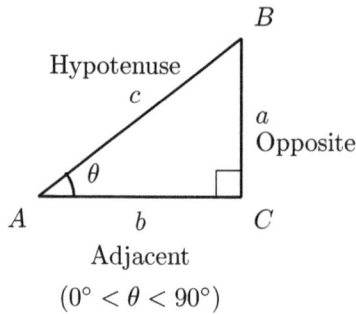

$$c^2 = a^2 + b^2, \qquad c = \sqrt{a^2 + b^2},$$
$$a^2 = c^2 - b^2, \qquad a = \sqrt{c^2 - b^2},$$
$$b^2 = c^2 - a^2, \qquad b = \sqrt{c^2 - a^2}.$$

Given the measure of an angle θ, we name three sides of the triangle as hypotenuse, opposite side, and adjacent side in reference to the relative position of sides to the angle as in the picture above. Six trigonometric values of the angle (**sine, cosine, tangent, cosecant, secant,** and **cotangent**) are defined as the ratios of the sides. The precise definitions are as follows.

✏️ The Trigonometry of Right Triangles

$$\sin\theta = \frac{\text{Opp}}{\text{Hyp}} = \frac{a}{c} \qquad\qquad \csc\theta = \frac{\text{Hyp}}{\text{Opp}} = \frac{c}{a}$$

$$\cos\theta = \frac{\text{Adj}}{\text{Hyp}} = \frac{b}{c} \qquad\qquad \sec\theta = \frac{\text{Hyp}}{\text{Adj}} = \frac{c}{b}$$

$$\tan\theta = \frac{\text{Opp}}{\text{Adj}} = \frac{a}{b} \qquad\qquad \cot\theta = \frac{\text{Adj}}{\text{Opp}} = \frac{b}{a}$$

Mnemonic: SOH   CAH   TOA

Example 1. Finding Trigonometric Function Values Given a Right Triangle.

Find the six trigonometric function values of the angle θ in the figure.

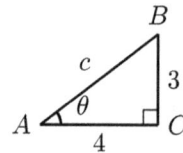

Solution.

Step 1. Find the hypotenuse. By the Pythagorean theorem, $c^2 = 3^2 + 4^2 = 9 + 16 = 25$. So, by taking square roots, $c = \pm\sqrt{25} = \pm 5$. The lengths of sides must be positive. Thus, $c = 5$.

Step 2. Label the sides of the triangle with values $a = 4, b = 3$, and $c = 5$.

Step 3. Find the values of trigonometric functions as ratios.

$$\sin\theta = \frac{\text{Opp}}{\text{Hyp}} = \frac{4}{5} \qquad\qquad\qquad \csc\theta = \frac{\text{Hyp}}{\text{Opp}} = \frac{5}{4}$$

$$\cos\theta = \frac{\text{Adj}}{\text{Hyp}} = \frac{3}{5} \qquad\qquad \sec\theta = \frac{\text{Hyp}}{\text{Adj}} = \frac{5}{3}$$

$$\tan\theta = \frac{\text{Opp}}{\text{Adj}} = \frac{4}{3} \qquad\qquad \cot\theta = \frac{\text{Adj}}{\text{Opp}} = \frac{3}{4}$$

Practice Now. Find the exact values of six trigonometric functions of an acute angle θ in a right triangle, assuming that the lengths of the side opposite and the side adjacent to θ are 7 and 24, respectively.

Example 2. Finding Other Trigonometric Function Values Given One

If for an acute angle α, $\sin\alpha = \dfrac{3}{4}$, find the values of five other trigonometric functions of α.

Solution. We first sketch a right triangle with an acute angle α. Because $\sin\alpha$ is defined as the ratio of the opposite side to the hypotenuse, we first sketch a right triangle with acute angle α, hypotenuse of length 4, and a side of length 3 opposite to α. If the adjacent side is x, then by the Pythagorean theorem, $3^2 + x^2 = 4^2$ or $x^2 = 7$, so $x = \sqrt{7}$.

$$\sin\alpha = \frac{\text{Opp}}{\text{Hyp}} = \frac{3}{4} \qquad\qquad \csc\alpha = \frac{\text{Hyp}}{\text{Opp}} = \frac{4}{3}$$

$$\cos\alpha = \frac{\text{Adj}}{\text{Hyp}} = \frac{\sqrt{7}}{4} \qquad\qquad \sec\alpha = \frac{\text{Hyp}}{\text{Adj}} = \frac{4}{\sqrt{7}}$$

$$\tan\alpha = \frac{\text{Opp}}{\text{Adj}} = \frac{3}{\sqrt{7}} \qquad\qquad \cot\alpha = \frac{\text{Adj}}{\text{Opp}} = \frac{\sqrt{7}}{3}$$

Practice Now. If for an acute angle α, $\cos\alpha = \dfrac{2}{7}$, find the other five trigonometric functions of α.

▶ **Special Right Triangles**

Trigonometric function values of 30°, 45°, and 60° are obtained from two special right triangles. The first one is an equilateral triangle folded in halves, and the second is an isosceles right triangle.

30°-60°-90° triangle   45°-45°-90° triangle

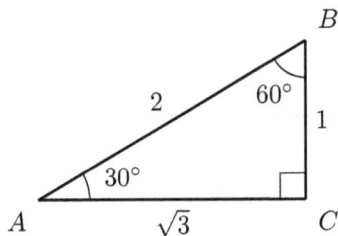 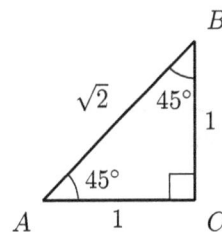

$\theta°$	θ rad	$\sin\theta$	$\cos\theta$	$\tan\theta$	$\csc\theta$	$\sec\theta$	$\cot\theta$
$30°$	$\dfrac{\pi}{6}$	$\dfrac{1}{2}$	$\dfrac{\sqrt{3}}{2}$	$\dfrac{\sqrt{3}}{3}$	2	$\dfrac{2\sqrt{3}}{3}$	$\sqrt{3}$
$45°$	$\dfrac{\pi}{4}$	$\dfrac{\sqrt{2}}{2}$	$\dfrac{\sqrt{2}}{2}$	1	$\sqrt{2}$	$\sqrt{2}$	1
$60°$	$\dfrac{\pi}{3}$	$\dfrac{\sqrt{3}}{2}$	$\dfrac{1}{2}$	$\sqrt{3}$	$\dfrac{2\sqrt{3}}{3}$	2	$\dfrac{\sqrt{3}}{3}$

Example 3. Using Special Trigonometric Values

Find the exact values of the following expressions.

a) $\sin 45° \cos 30° - \cos 45° \sin 30°$

b) $\sin^2 \dfrac{\pi}{3} + \tan \dfrac{\pi}{4}$

Solution.   Use the values in the table. Note that $\sin^2 \dfrac{\pi}{3}$ is a short-hand notation for $\left(\sin \dfrac{\pi}{3}\right)^2$.

a) $\sin 45° \cos 30° - \cos 45° \sin 30°$

$= \dfrac{\sqrt{2}}{2} \cdot \dfrac{\sqrt{3}}{2} - \dfrac{\sqrt{2}}{2} \cdot \dfrac{1}{2} = \dfrac{\sqrt{6}-\sqrt{2}}{4}$

b) $\sin^2 \dfrac{\pi}{3} + \tan \dfrac{\pi}{4}$

$= \left(\dfrac{\sqrt{3}}{2}\right)^2 + 1 = \dfrac{3}{4} + 1 = \dfrac{7}{4}$

Practice Now.   Find the exact values of the following expressions.

a) $\sin^2 30° + \cos^2 30°$

b) $\sec \dfrac{\pi}{6} + \cot \dfrac{\pi}{3}$

▶ Cofunction Identities for Complementary Angles

A trigonometric function of an angle is always equal to the cofunction of the complement of the angle.

✏ **Cofunction Identities**

$$\sin\theta = \cos\left(90° - \theta\right) \qquad\qquad \cos\theta = \sin\left(90° - \theta\right)$$
$$\tan\theta = \cot\left(90° - \theta\right) \qquad\qquad \cot\theta = \tan\left(90° - \theta\right)$$
$$\sec\theta = \csc\left(90° - \theta\right) \qquad\qquad \csc\theta = \sec\left(90° - \theta\right)$$

If θ is measured in radians, replace $90°$ with $\dfrac{\pi}{2}$ in the equations above.

Example 4. Using Cofunction Identities

a) Given $\tan 22° \approx 0.4040$, find $\cot 68°$.

b)   Given $\sin 18° \approx 0.3090$, find $\cos 72°$.

Solution.

a) Because $22° = 90° - 68°$, the complement of $22°$ angle is $62°$. So,

$$\cot 68° = \cot\left(90° - 22°\right) = \tan 22° \approx 0.4040.$$

b) Because $72° = 90° - 18°$, the complement of $18°$ angle is $72°$. So,

$$\cos 72° = \cos\left(90° - 18°\right) = \sin 18° \approx 0.3090.$$

Practice Now.

a) Given $\sec 69° \approx 2.7904$, find $\csc 21°$.   **b)** Given that $\cot 15° \approx 3.7321$, find $\tan 75°$.

▶ **Fundamental Identities**

Fundamental identities describe fundamental relations between six trigonometric functions. The equality holds for any value of θ. They can be used to find all trigonometric function values given one of them.

🖉 **Fundamental Identities**

- Reciprocal Identities

$$\sin\theta = \frac{1}{\csc\theta} \qquad \cos\theta = \frac{1}{\sec\theta} \qquad \tan\theta = \frac{1}{\cot\theta}$$

$$\csc\theta = \frac{1}{\sin\theta} \qquad \sec\theta = \frac{1}{\cos\theta} \qquad \cot\theta = \frac{1}{\tan\theta}$$

- Quotient Identities

$$\tan\theta = \frac{\sin\theta}{\cos\theta} \qquad \cot\theta = \frac{\cos\theta}{\sin\theta}$$

- Pythagorean Identities

$$\sin^2\theta + \cos^2\theta = 1 \qquad 1 + \tan^2\theta = \sec^2\theta \qquad 1 + \cot^2\theta = \csc^2\theta$$

Example 5. Finding Trigonometric Function Values Using Fundamental Identities

Let θ be an acute angle with $\cos\theta = \dfrac{3}{5}$. Find the following.

a) $\sin\theta$ **b)** $\sec\theta$ **c)** $\tan\theta$ **d)** $\csc(90° - \theta)$

Solution.

a) Because $\sin^2\theta + \cos^2\theta = 1$, by substituting the cosine value, we get $\sin^2\theta + \left(\dfrac{3}{5}\right)^2 = 1$. Solve the

equation for $\sin\theta$. Subtract the squared fraction from both sides to get $\sin^2\theta = 1 - \left(\dfrac{3}{5}\right)^2 = \dfrac{16}{25}$, and

take square roots to get the answer $\sin\theta = \sqrt{\dfrac{16}{25}} = \dfrac{4}{5}$.

b) Apply a reciprocal identity; then $\sec\theta = \dfrac{1}{\cos\theta} = \dfrac{1}{\frac{3}{5}} = \dfrac{5}{3}$.

c) Apply a quotient identity; then $\tan\theta = \dfrac{\sin\theta}{\cos\theta} = \dfrac{\frac{4}{5}}{\frac{3}{5}} = \dfrac{4}{3}$.

d) Apply a cofunction identity; then $\csc\left(90° - \theta\right) = \sec\theta = \dfrac{5}{3}$.

Practice Now.   Let θ be an acute angle with $\sin\theta = \dfrac{12}{13}$. Find the following.

a) $\cos\theta$ **b)** $\csc\theta$ **c)** $\tan\theta$ **d)** $\cot(90° - \theta)$

Exercises 2.2

(1–6) Use the figure below and the given values to find each specified trigonometric function value. Rationalize denominators where necessary.

1. $a = 8$, $b = 10$ Find $\sin\theta$ and $\cot\theta$.

2. $b = 4$, $c = 12$ Find $\cos\theta$ and $\csc\theta$.

3. $a = 4$, $c = 7$ Find $\cos\theta$ and $\tan\theta$.

4. $a = 3$, $b = 5$ Find $\cos\theta$. and $\sin\theta$.

5. $a = 7$, $b = 24$ Find $\sec\theta$ and $\tan\theta$.

6. $a = 2$, $c = 3$ Find $\cot\theta$ and $\csc\theta$.

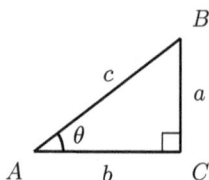

(7–12) Suppose θ is an acute angle. Use the given value to find five other trigonometric function values of θ by sketching a right triangle.

7. $\sin\theta = \dfrac{5}{13}$ **8.** $\cos\theta = \dfrac{9}{40}$

9. $\tan\theta = \sqrt{3}$ **10.** $\sec\theta = \dfrac{7}{2}$

11. $\csc\theta = \dfrac{3}{2}$ **12.** $\cot\theta = 1$

(13–18) Use the table of trigonometric ratios for special angles to evaluate the expression.

13. $\sin\dfrac{\pi}{6} + \cos\dfrac{\pi}{6}$

14. $\sin 30° \csc 30°$

15. $\sin 45° \cos 60° + \cos 45° \sin 60°$

16. $\cos^2 30° + \sin^2 30°$

17. $\left(\sin\dfrac{\pi}{3}\cos\dfrac{\pi}{4} - \sin\dfrac{\pi}{4}\cos\dfrac{\pi}{3}\right)^2$

18. $\tan 60° \sec 45° + \tan 30° \cos 60°$

(19–24) Use a cofunction identity to find the trigonometric function value of the complementary angle.

19. Given that $\sin 52° \approx 0.7880$, find $\cos 38°$.

20. Given that $\cos 30° \approx 0.8660$, find $\sin 60°$.

21. Given that $\tan 23° \approx 0.4245$, find $\cot 67°$.

22. Given that $\csc 25° \approx 2.3662$, find $\sec 65°$.

23. Given that $\sec 12° \approx 1.0223$, find $\csc 78°$.

24. Given that $\cot 63° \approx 0.5095$, find $\tan 27°$.

(25–30) Use the given value to find the indicated trigonometric function values of the acute angle θ.

25. $\cos\theta = \dfrac{7}{12}$; find $\sin\theta$, $\sec\theta$, $\tan\theta$, and $\csc(90° - \theta)$.

26. $\sin\theta = \dfrac{4}{7}$; find $\cos\theta$, $\sec\theta$, $\tan\theta$, and $\cot(90° - \theta)$

27. $\csc\theta = \dfrac{3}{2}$; find $\sin\theta$, $\cos\theta$, $\tan\theta$, and $\sec(90° - \theta)$

28. $\tan\theta = \dfrac{5}{3}$; find $\sin\theta$, $\cos\theta$, $\sec\theta$, and $\cot(90° - \theta)$

29. $\sec\theta = 3$; find $\sin\theta$, $\cos\theta$, $\tan\theta$, and $\cot(90° - \theta)$

30. $\cot\theta = \dfrac{5}{3}$; find $\sin\theta$, $\tan\theta$, $\sec\theta$, and $\sin(90° - \theta)$

2.3 Trigonometric Functions of Any Angle

▶ **The Definition of Trigonometric Functions of Any Angle**

Let (x, y) be a point other than the origin on the terminal side of an angle θ in standard position. Let r be the distance between the point (x, y) and the origin. Then x, y, and r are related by the distance formula.

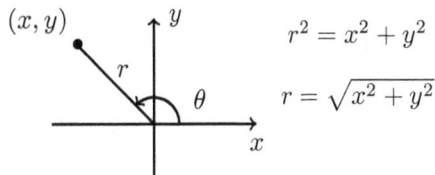

$$r^2 = x^2 + y^2$$

$$r = \sqrt{x^2 + y^2}$$

Six trigonometric functions are defined to be the ratios of pairs of x, y, and r as follows.

✎ **Trigonometric Functions of Any Angle**

$$\sin\theta = \frac{y}{r} \qquad\qquad \csc\theta = \frac{r}{y} \quad (y \neq 0)$$

$$\cos\theta = \frac{x}{r} \qquad\qquad \sec\theta = \frac{r}{x} \quad (x \neq 0)$$

$$\tan\theta = \frac{y}{x} \quad (x \neq 0) \qquad\qquad \cot\theta = \frac{x}{y} \quad (y \neq 0)$$

The definition is consistent with the definition in the previous section in case θ is acute.

▶ **The Signs of Trigonometric Functions**

The signs of trigonometric functions are determined by the quadrant in which the angle θ lies. For example, if θ is in quadrant II, then $x < 0$, $y > 0$, and $r > 0$. Thus, $\sin\theta = \dfrac{y}{r} > 0$ and $\cos\theta = \dfrac{x}{r} < 0$, and so on. The following chart lists the signs of all trigonometric functions in all four quadrants. An easy way to remember these signs is to remember what functions are positive in each quadrant. Use the mnemonic "All(I) Students(II) Take(III) Calculus(IV)" to remember them.

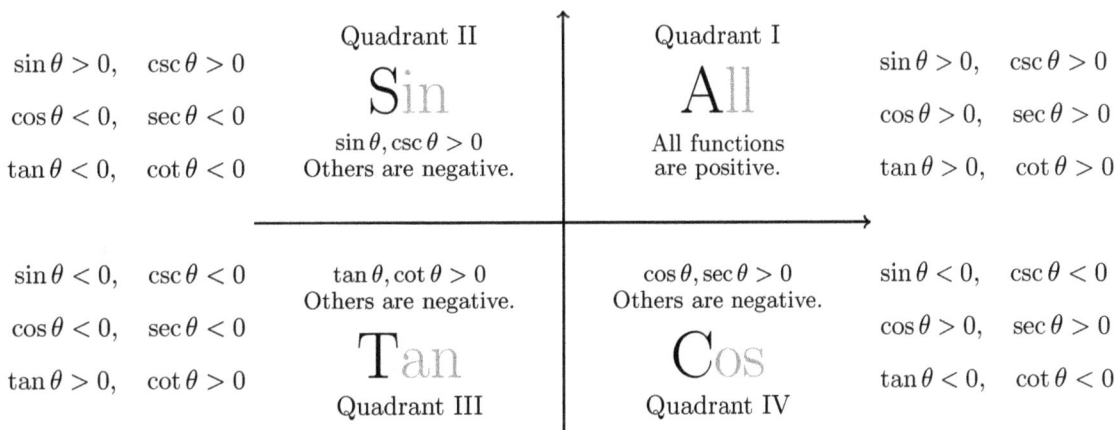

	Quadrant II	Quadrant I	
$\sin\theta > 0, \quad \csc\theta > 0$	**Sin**	**All**	$\sin\theta > 0, \quad \csc\theta > 0$
$\cos\theta < 0, \quad \sec\theta < 0$			$\cos\theta > 0, \quad \sec\theta > 0$
$\tan\theta < 0, \quad \cot\theta < 0$	$\sin\theta, \csc\theta > 0$ Others are negative.	All functions are positive.	$\tan\theta > 0, \quad \cot\theta > 0$
$\sin\theta < 0, \quad \csc\theta < 0$	$\tan\theta, \cot\theta > 0$ Others are negative.	$\cos\theta, \sec\theta > 0$ Others are negative.	$\sin\theta < 0, \quad \csc\theta < 0$
$\cos\theta < 0, \quad \sec\theta < 0$			$\cos\theta > 0, \quad \sec\theta > 0$
$\tan\theta > 0, \quad \cot\theta > 0$	**Tan** Quadrant III	**Cos** Quadrant IV	$\tan\theta < 0, \quad \cot\theta < 0$

Example 1. Finding the Quadrant in which the Angle Lies Using Signs of Trigonometric Functions

In which quadrant does the angle θ lie if $\tan \theta < 0$ and $\cos \theta < 0$?

Solution. Because $\tan \theta < 0$, θ lies in quadrant II or IV. Also, given that $\cos \theta < 0$, θ lies in quadrant II or III. The quadrant satisfying both conditions is quadrant II.

Practice Now. In which quadrant does the angle θ lie if $\sin \theta < 0$ and $\cos \theta > 0$?

Example 2. Finding Trigonometric Function Values Given a Trigonometric Function Value and a Sign

Given that $\tan \theta = -\dfrac{3}{4}$ and $\cos \theta < 0$, find $\cos \theta$ and $\sin \theta$.

Solution. Because the tangent is negative and the cosine is negative, θ lies in quadrant II. In quadrant II, x is negative and y is positive. From the definition of the trigonometric functions,

$$\tan \theta = -\frac{3}{4} = \frac{3}{-4} = \frac{y}{x}.$$

Thus, we may assume $y = 3$ and $x = -4$. Then,

$$r = \sqrt{x^2 + y^2} = \sqrt{(-4)^2 + 3^2} = \sqrt{16 + 9} = 5.$$

Therefore, $\cos \theta = \dfrac{x}{r} = \dfrac{-4}{5} = -\dfrac{4}{5}$, and $\sin \theta = \dfrac{y}{r} = \dfrac{3}{5}$.

Practice Now. Given that $\tan \theta = -\dfrac{2}{3}$ and $\cos \theta < 0$, find $\sin \theta$ and $\cos \theta$.

▶ Trigonometric Function Values of Coterminal Angles

Two angles in standard position with the same terminal side are called **coterminal angles**. Two angles are coterminal if and only if the difference between them is a multiple of $360°$ or 2π in radian. So, any angle coterminal to an angle θ can be written as $\theta + 360°k$ or $\theta + 2\pi k$ for an integer k, and vice versa. Because only the terminal side matters for trigonometric functions, coterminal angles have the same trigonometric function values.

✎ **Trigonometric Function Values of Coterminal Angles**

For any integer k,

$$\sin \theta = \sin(\theta + 360°k), \qquad \cos \theta = \cos(\theta + 360°k),$$
$$\sin \theta = \sin(\theta + 2\pi k), \qquad \cos \theta = \cos(\theta + 2\pi k).$$

===

Example 3. Finding a Coterminal Angle

Find the angle between $0°$ and $360°$ that is coterminal to the given angle.

a) $1180°$ **b)** $-430°$

===

Solution. We can get coterminal angles by repeatedly adding or subtracting $360°$.

a) Because $1180°$ is greater than $360°$, subtract $360°$ repeatedly until you get an angle small enough.

$$1180° - 360° = 820°, \qquad 820° - 360° = 460°, \qquad 460° - 360° = 100°.$$

Or, equivalently, divide $1180°$ by $360°$ to get the remainder.

$$\frac{1180°}{360°} = 3 + \frac{100°}{360°}$$

Therefore, the answer is $100°$.

b) Because $-430°$ is less than $0°$, add $360°$ repeatedly to get a positive angle.

$$-430° + 360° = -70°, \qquad -70° + 360° = 290°.$$

The answer is $290°$.

Practice Now. Find the angle between $0°$ and $360°$ that is coterminal to the given angle.

a) $510°$ **b)** $-940°$

▶ **Reference Angles**

Let θ be a non-quadrantal angle in standard position, that is, an angle whose terminal side does not fall on one of coordinate axes. Its **reference angle**, θ', is the positive acute angle measured counterclockwise between the positive or negative x-axis (whichever is closer) and the terminal side of θ. To determine the reference angle, find a coterminal angle between $0°$ and $360°$, then use one of the following formulas.

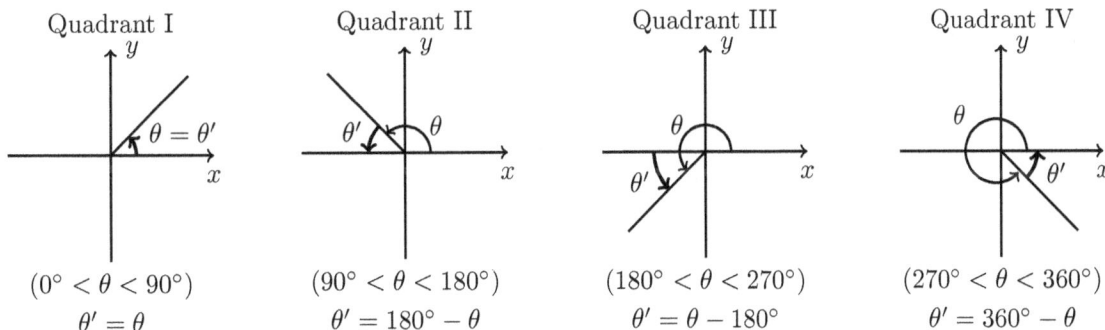

Quadrant I

$(0° < \theta < 90°)$
$\theta' = \theta$

Quadrant II

$(90° < \theta < 180°)$
$\theta' = 180° - \theta$

Quadrant III

$(180° < \theta < 270°)$
$\theta' = \theta - 180°$

Quadrant IV

$(270° < \theta < 360°)$
$\theta' = 360° - \theta$

===

Example 4. Finding the Reference Angle

Find the reference angle of the given angle.

a) $690°$ **b)** $\dfrac{5\pi}{7}$

===

Solution.   The method is to reduce the angle as much as possible by taking the difference with $180°$ (π in radians) or $360°$ (2π in radians).

a) Subtract $360°$ from the angle to get $330°$. Because the terminal side of the angle $330°$ in standard position lies in quadrant IV, the reference angle is

$$\theta' = 360° - 330° = 30°.$$

b) Because the terminal side of an angle $\dfrac{5\pi}{7}$ lies in quadrant II, the reference angle is

$$\theta' = \pi - \frac{5\pi}{7} = \frac{2\pi}{7}.$$

Practice Now.   Find the reference angle of the given angle.

a) $520°$

b) $\dfrac{11\pi}{12}$

The trigonometric function values of θ and those of the reference angle θ' are either equal or opposite to each other. In other words,

$$\sin\theta = \pm\sin\theta', \qquad \cos\theta = \pm\cos\theta', \qquad \tan\theta = \pm\tan\theta',$$
$$\csc\theta = \pm\csc\theta', \qquad \sec\theta = \pm\sec\theta', \qquad \cot\theta = \pm\cot\theta'.$$

The sign is determined by the quadrant in which the angle θ lies.

Example 5. Finding the Exact Trigonometric Function Value Using the Reference Angle

Find the exact value of $\sin 585°$.

Solution.   Subtract $585° - 360° = 225°$ to get a coterminal angle. Because the terminal side of the angle lies in quadrant III, the reference angle of $225°$ is $225° - 180° = 45°$. The sine value is negative in quadrant III. Thus,

$$\sin 585° = -\sin 45° = -\frac{\sqrt{2}}{2}.$$

Practice Now.   Find the exact value of the trigonometric function.

a) $\cos 690°$

b) $\tan(-150°)$

▶ Unit Circle Trigonometry

The unit circle is the circle with radius 1, that is, $r = 1$, centered at the origin. Let $P(x,y)$ be the point on the unit circle that corresponds to an angle θ, then the point satisfies the unit circle equation $x^2 + y^2 = 1$, and $\cos\theta = \dfrac{x}{1} = x$ and $\sin\theta = \dfrac{y}{1} = y$.

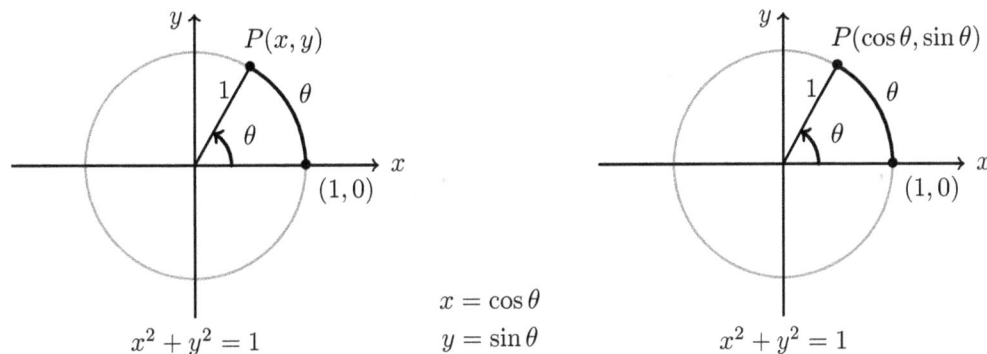

$$x = \cos\theta$$
$$y = \sin\theta$$

Trigonometric Functions: The Unit Circle Approach

Trigonometric function values of θ and the coordinates of the point (x, y) corresponding to the angle on the unit circle are related as follows.

$$\sin\theta = y \qquad\qquad \csc\theta = \frac{1}{y} \quad (y \neq 0)$$

$$\cos\theta = x \qquad\qquad \sec\theta = \frac{1}{x} \quad (x \neq 0)$$

$$\tan\theta = \frac{y}{x} \quad (x \neq 0) \qquad\qquad \cot\theta = \frac{x}{y} \quad (y \neq 0)$$

The unit circle chart in Figure 2.1 lists special trigonometric values. The values in the first quadrant were obtained in Section 2.2. The values in other quadrants are obtained using the symmetry.

Example 6. Finding Trigonometric Function Values Using the Unit Circle

Find the exact values of the following expressions.

a) $\cos 180°$ **b)** $\sin 405°$ **c)** $\tan 120°$ **d)** $\csc\left(-\frac{13\pi}{4}\right)$

Solution. Remember that the x-coordinate represents the cosine value and the y-coordinate represents the sine value in the unit circle.

a) The terminal side of the angle 180° intersects the unit circle at the point $(-1, 0)$. Using the property $\cos\theta = x$, we conclude that $\cos 180° = -1$.

b) Reduce the angle by subtracting 360° to get a coterminal angle $405° - 360° = 45°$. The terminal side of the angle 45° intersects the unit circle at the point $\left(\frac{\sqrt{2}}{2}, \frac{\sqrt{2}}{2}\right)$. Using the property $\sin\theta = y$, we conclude that $\sin 45° = \frac{\sqrt{2}}{2}$.

c) The terminal side of the 120° intersects the unit circle at the point $\left(-\frac{1}{2}, \frac{\sqrt{3}}{2}\right)$. Using the property

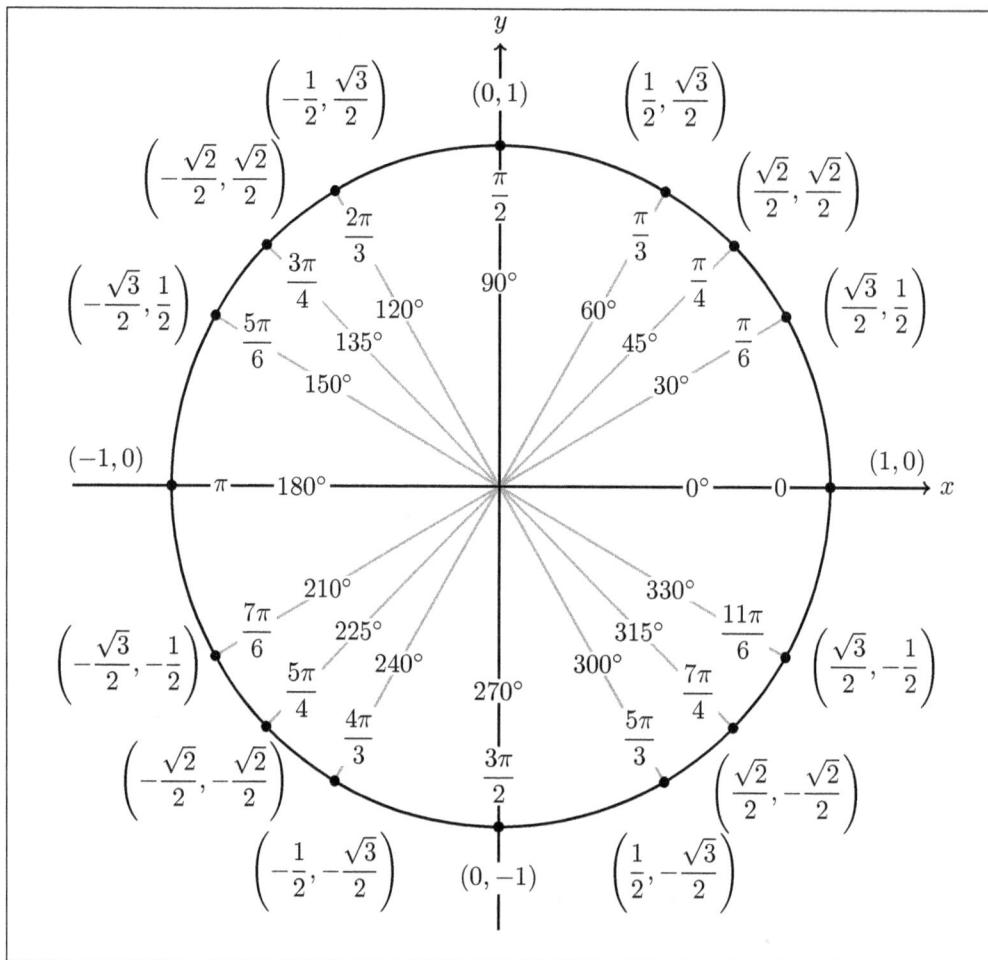

Figure 2.1. The Unit Circle Chart

$\tan\theta = \dfrac{y}{x}$, we conclude that

$$\tan 120° = \frac{\dfrac{\sqrt{3}}{2}}{-\dfrac{1}{2}} = \frac{\sqrt{3}}{2} \times \left(-\frac{2}{1}\right) = -\sqrt{3}.$$

d) Simplify the angle by repeatedly adding 2π.

$$-\frac{11\pi}{4} + 2\pi = -\frac{3\pi}{4}, \qquad -\frac{3\pi}{4} + 2\pi = \frac{5\pi}{4}.$$

The terminal side of the angle $\dfrac{5\pi}{4}$ intersects the unit circle at the point $\left(-\dfrac{\sqrt{2}}{2}, -\dfrac{\sqrt{2}}{2}\right)$. Using the definition $\csc\theta = \dfrac{1}{\sin\theta} = \dfrac{1}{y}$, we conclude that

$$\csc 225° = \frac{1}{-\dfrac{\sqrt{2}}{2}} = -\frac{2}{\sqrt{2}} = -\frac{2}{\sqrt{2}} \cdot \frac{\sqrt{2}}{\sqrt{2}} = -\frac{2\sqrt{2}}{2} = -\sqrt{2}.$$

Practice Now.   Find the exact values of the following expressions.

a) $\cos 270°$ **b)** $\sin 150°$ **c)** $\tan 690°$ **d)** $\sec(-330°)$

Exercises 2.3

(1–6) The terminal side of angle θ in standard position contains the given point. Find the exact values of the six trigonometric funcions of θ.

1. $(1,3)$ **2.** $(-5, 12)$

3. $(-3, -4)$ **4.** $(7, -12)$

5. $(-\sqrt{3}, 1)$ **6.** $(4, 4)$

(7–12) Find the quadrant in which θ lies from the given information.

7. $\tan \theta > 0$ and $\sin \theta < 0$ **8.** $\cos \theta < 0$ and $\csc \theta < 0$

9. $\sin \theta > 0$ and $\cos \theta < 0$ **10.** $\csc \theta > 0$ and $\sec \theta > 0$

11. $\cos \theta > 0$ and $\csc \theta < 0$ **12.** $\tan \theta < 0$ and $\sin \theta < 0$

(13–18) Find the exact values of the remaining trigonometric functions of θ from the given information.

13. $\sin \theta = -\dfrac{5}{13}$, θ in quadrant III

14. $\cos \theta = \dfrac{12}{13}$, θ in quadrant IV

15. $\tan \theta = -\dfrac{3}{4}$, θ in quadrant II

16. $\sec \theta = -\dfrac{25}{24}$, $\tan \theta > 0$

17. $\cot \theta = \dfrac{3}{7}$, $\sec \theta > 0$

18. $\tan \theta = -5$, $\sin \theta < 0$

(19–25) Find the reference angle of the given angle.

19. $150°$ **20.** $330°$

21. $2115°$ **22.** $-290°$

23. $\dfrac{4\pi}{3}$ **24.** $\dfrac{5\pi}{7}$

25. $\dfrac{33\pi}{4}$

(26–33) Find the exact value of the trigonometric function.

26. $\sin 2115°$ **27.** $\sin(-1320°)$

28. $\tan 690°$ **29.** $\cos 1920°$

30. $\sec(-2490°)$ **31.** $\csc 2640°$

32. $\sec \dfrac{43\pi}{4}$ **33.** $\cos\left(-\dfrac{32\pi}{3}\right)$

2.4   Graphs of Sine and Cosine Functions

▶ **Periodicity of Trigonometric Functions**

A function f is said to be periodic with period p (p being a nonzero constant) if we have

$$f(x + p) = f(x)$$

for all values of x in the domain of f, and p is the smallest such value. A function with period p will repeat on intervals of length p, and these intervals are referred to as **periods**. The six trigonometric functions are periodic, and the graphs of these functions have patterns repeating indefinitely. The patterns repeat every period.

📓 **Periodicity of Sine, Cosine, and Tangent**

The sine and cosine functions are periodic with period 2π.

$$\sin(x + 2\pi) = \sin x, \qquad \cos(x + 2\pi) = \cos x$$

The tangent function is periodic with period π.

$$\tan(x + \pi) = \tan x$$

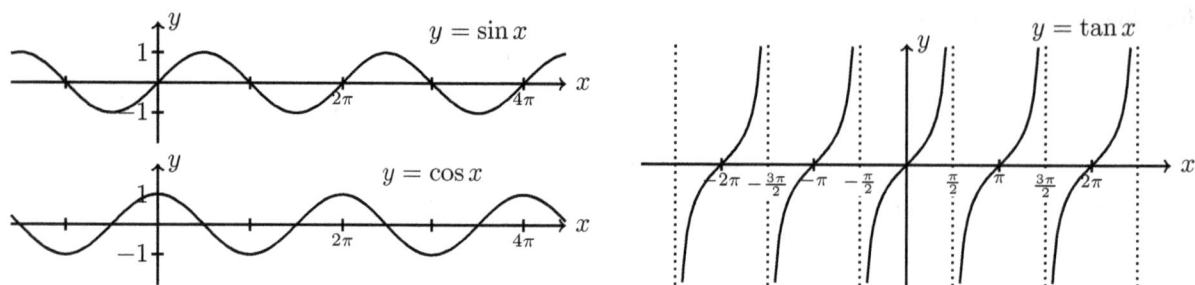

Figure 2.2. Graphs of Sine, Cosine, and Tangent

Example 1. Using Periodicity to Find Trigonometric Function Values

Find the value of the trigonometric function.

a) $\sin \dfrac{13\pi}{6}$

b) $\tan \dfrac{5\pi}{3}$

Solution.   Reduce the angle first using periods of trigonometric functions.

a) Because the function sine is periodic with period 2π, we have $\sin(x + 2\pi) = \sin x$. Therefore,

$$\sin \frac{13\pi}{6} = \sin\left(\frac{\pi}{6} + 2\pi\right) = \sin \frac{\pi}{6} = \frac{1}{2}.$$

b) Because the function tangent is periodic with period π, we have $\tan(x + \pi) = \tan x$. Therefore,

$$\tan \frac{5\pi}{3} = \tan\left(\frac{2\pi}{3} + \pi\right) = \tan \frac{2\pi}{3} = -\sqrt{3}.$$

Practice Now.   Find the value of the trigonometric function.

a) $\cos \dfrac{11\pi}{3}$

b) $\tan \dfrac{7\pi}{4}$

▶ **Graphing Transformations of Sine and Cosine Functions**

We will study the properties of the graphs of sine and cosine functions, and learn how to draw transformations of these graphs. Let's start with the basic sine and cosine functions, $\sin x$ and $\cos x$. These functions have an amplitude of 1. The amplitude is the amount of fluctuation of a wave from the midline. The amplitude of the functions is 1 because the graph goes one unit up and one unit down from the midline of the graph. These functions have a period of 2π; thus, the sine and cosine wave repeats every 2π units. The following graphs show sine and cosine patterns over the interval $[0, 2\pi]$.

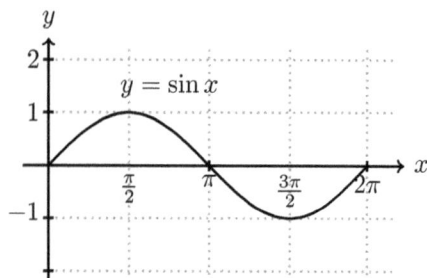

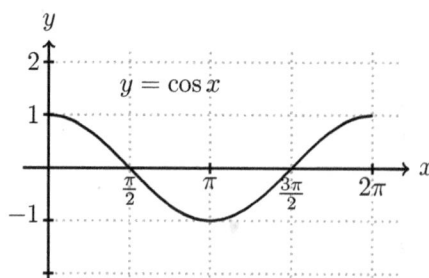

We get a transformed graph through scaling and shifting of the basic graph. The amplitude, period, and position change as a result of the transformation.

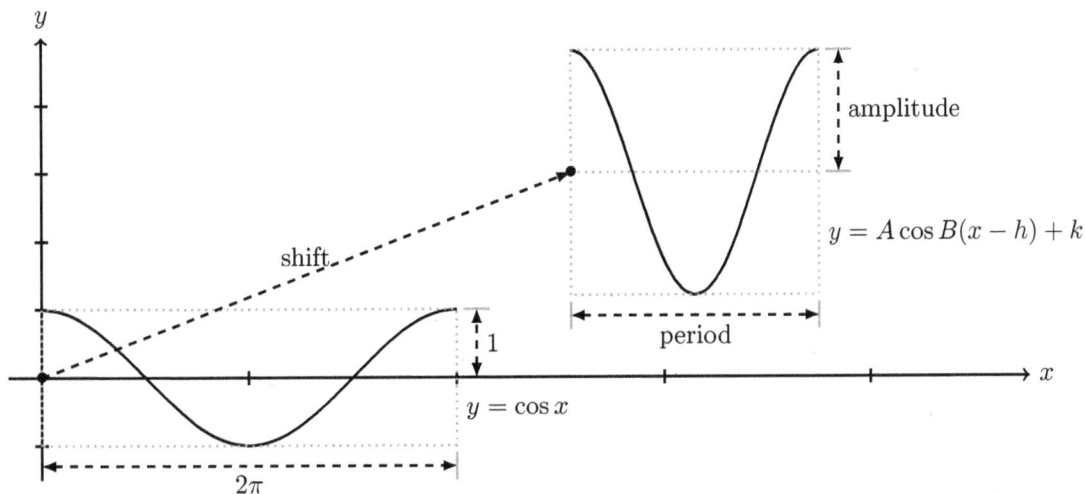

Figure 2.3. A Transformation of Cosine Graph

Transformations of the sine and cosine functions can be written in standard form as follows.

$$y = A \sin B(x - h) + k, \qquad y = A \cos B(x - h) + k,$$

where A, B, h, and k are constants and $B > 0$. The constant A is related to the amplitude, B is related to the period, and h and k are phase (horizontal) and vertical shifts, respectively.

✏️ **Properties of Transformed Sine and Cosine Graphs**

For $y = A \sin B(x - h) + k$ or $y = A \cos B(x - h) + k$, the graph has the following properties.

- amplitude $= |A|$, (if $A < 0$, the graph is turned upside down.)

- period $= \dfrac{2\pi}{B}$

- phase shift $= h$

- vertical shift $= k$

Example 2. Graphing Sine or Cosine with Different Amplitude and Period

Graph the function $y = \dfrac{1}{2} \cos 3x$ over the interval $[0, 2\pi)$.

Solution.   By comparision we can get that $A = \dfrac{1}{2}$ and $B = 3$; therefore, the period of the function is

$\frac{2\pi}{B} = \frac{2\pi}{3}$, and the amplitude is $\frac{1}{2}$. Draw the cosine pattern over one period and repeat the pattern. The graph is as follows.

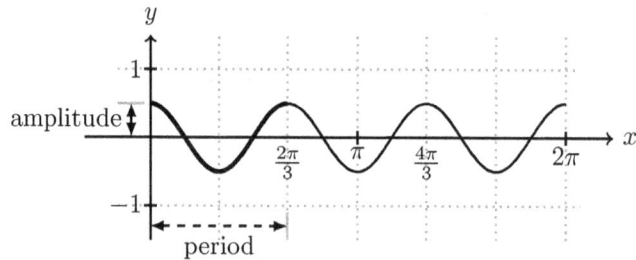

Practice Now.   Graph the function $y = 2\sin 4x$ over the interval $[0, 2\pi)$.

Example 3. Graphing Sine or Cosine with Different Amplitude and Phase Shift

Find the amplitude, period, and phase shift, then sketch the graph of $y = -\frac{3}{2}\sin\left(x + \frac{\pi}{2}\right)$.

Solution.   Identify the values of $A = -\frac{3}{2}$, $B = 1$, $h = -\frac{\pi}{2}$, and $k = 0$. Then the amplitude is $\left|-\frac{3}{2}\right| = \frac{3}{2}$, the period is 2π, the phase shift is $-\frac{\pi}{2}$, and there is no vertical shift. Plot the anchor point at $(h, k) = \left(-\frac{\pi}{2}, 0\right)$, measure the period and amplitude from the point, then draw the sine pattern upside down (because $A < 0$) over one period. Repeat the pattern to draw the sine wave. The graph is as follows.

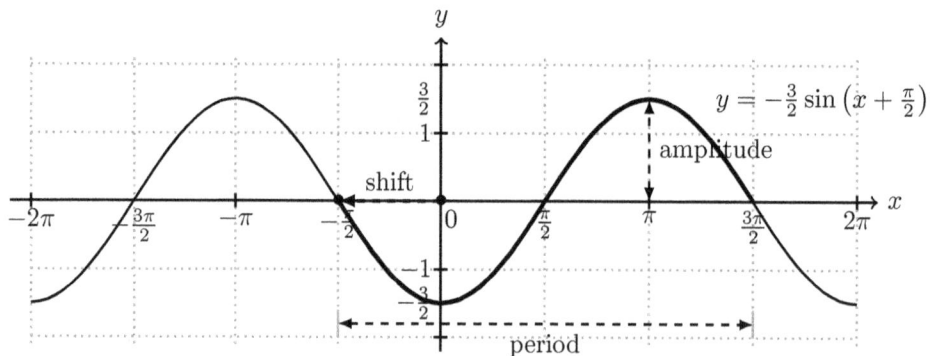

Practice Now.   Find the amplitude, period, and phase shift, then sketch the graph of $y = \frac{5}{2}\cos\left(x + \frac{\pi}{4}\right)$.

Example 4. Graphing Sine or Cosine with Different Amplitude, Period, Phase Shift, and Vertical Shift

Graph $y = 2\cos(2x - \pi) + 2$.

Solution.   Change the equation $y = 2\cos(2x - \pi) + 2$ to the standard form $y = 2\cos 2\left(x - \frac{\pi}{2}\right) + 2$ by factoring 2. Then identify the constants $A = 2$, $B = 2$, $h = \frac{\pi}{2}$, and $k = 2$. The amplitude is 2, the period is

$\dfrac{2\pi}{2} = \pi$, the phase shift is $\dfrac{\pi}{2}$, and the vertical shift up is 2 units. Plot the anchor point at $(h, k) = \left(\dfrac{\pi}{2}, 2\right)$, measure the period and amplitude from the point, then draw the cosine pattern over one period. Repeat the pattern to draw the cosine wave. The graph is as follows.

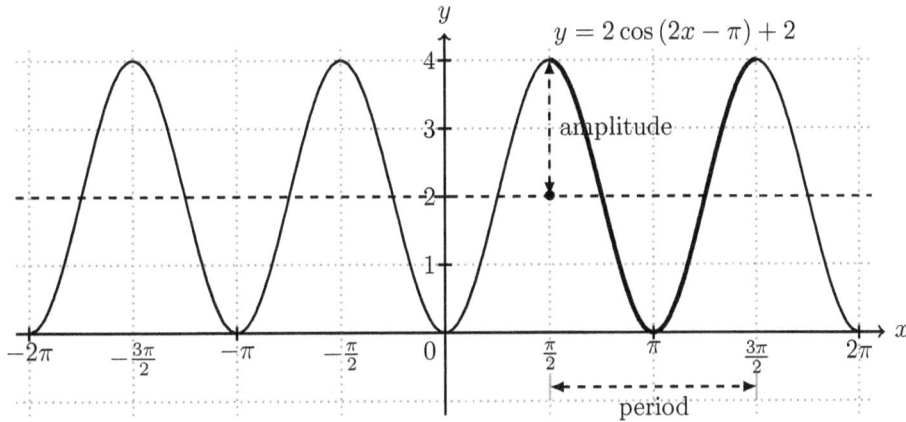

Practice Now.   Graph $y = \dfrac{3}{2} \sin\left(3x - \dfrac{\pi}{2}\right) + 1$.

▶ **Negative Angle Identities**

The graph of sine is symmetric with respect to the origin. On the other hand, the graph of cosine is symmetric with respect to the y-axis. We state these facts by saying that the sine function is odd and the cosine function is even. Algebraically, these properties are equivalent to

$$\sin(-x) = -\sin x, \quad \cos(-x) = \cos x \quad \text{for any } x.$$

The equations state that the sine function has the opposite value for the opposite angle, and the cosine function has the same value for the opposite angle.

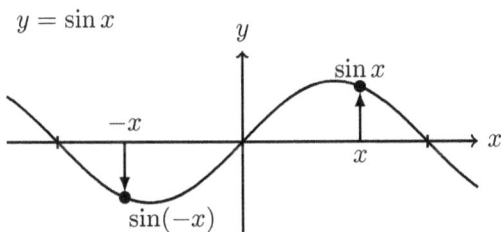

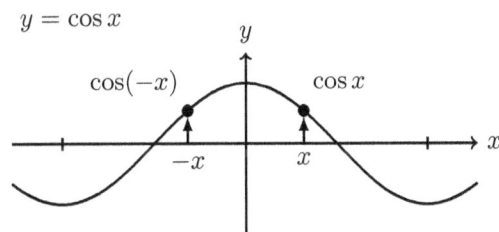

Example 5. Using Negative Angle Identities to Find the Properties of Sine and Cosine Functions

Find the amplitude, period, phase shift, and vertical shift of the trigonometric function.

a) $y = 3\sin(-x - \pi)$

b) $y = \cos\left(-\pi x + \dfrac{\pi}{3}\right) + 5$

Solution.   Factor the negative coefficient and use the negative angle identities for sine and cosine. Then find the stated properties.

a) Because $3\sin(-x-\pi) = 3\sin[-(x+\pi)] = -3\sin(x+\pi)$, the amplitude is $|-3| = 3$ (the graph is drawn upside down), the period is 2π, the phase shift is $-\pi$, and the vertical shift is 0.

b) Because $\cos\left(-\pi x + \dfrac{\pi}{3}\right) + 5 = \cos\left[-\pi\left(x - \dfrac{1}{3}\right)\right] + 5 = \cos\pi\left(x - \dfrac{1}{3}\right) + 5$, the amplitude is 1, the period is $\dfrac{2\pi}{\pi} = 2$, the phase shift is $\dfrac{1}{3}$, and the vertical shift is 5.

Practice Now. Find the amplitude, period, phase shift, and vertical shift of the trigonometric function.

a) $y = \dfrac{2}{3}\sin\left(-x + \dfrac{5\pi}{4}\right) + 3$ 　　　　　　　　　**b)** $y = -\cos\left(-\pi x - 3\pi\right) - 1$

Exercises 2.4

(1–8) Find the value of the following functions. Answer in fraction and radical form, not a decimal approximation.

1. $\sin\dfrac{11\pi}{4}$

2. $\cos\dfrac{7\pi}{3}$

3. $\tan\dfrac{7\pi}{4}$

4. $\cos\dfrac{9\pi}{4}$

5. $\tan\dfrac{4\pi}{3}$

6. $\sin\dfrac{23\pi}{4}$

7. $\cos\dfrac{11\pi}{6}$

8. $\tan\dfrac{11\pi}{3}$

(9–23) Determine the amplitude, period, phase shift, and vertical shift of each function.

9. $y = -2\sin 3x$

10. $y = 3\cos 2x$

11. $y = 3\sin\dfrac{2}{3}x$

12. $y = -4\cos 5x$

13. $y = 3\cos\dfrac{1}{2}x$

14. $y = \sin\left(x - \dfrac{\pi}{3}\right)$

15. $y = 2\cos x$

16. $y = -2\sin 2x$

17. $y = -\cos(-2x) - 5$

18. $y = \dfrac{1}{4}\sin(-2x)$

19. $y = -\sin(x - \pi)$

20. $y = 3\sin 6x - 3$

21. $y = 3\cos\dfrac{1}{2}(x - \pi) + 2$

22. $y = \cos 2\left(x - \dfrac{\pi}{3}\right)$

23. $y = -3.5\sin\left(2x - \dfrac{\pi}{2}\right) - 1$

24. Find an equation for a sine function that has an amplitude of 4 and a period of 3π.

25. Find an equation for a cosine function that has an amplitude of $\dfrac{3}{5}$ and a period of $\dfrac{3}{2}\pi$.

(26–32) Sketch the graph of the following functions.

26. $y = -3\cos(2\pi x + 4\pi)$

27. $y = 2\sin(2\pi x + 4\pi)$

28. $y = \sin x + 2$

29. $y = \cos(-2x - 2\pi)$

30. $y = 3\sin(-x) + 2$

31. $y = \cos(x + \pi) + 1$

32. $y = \dfrac{1}{2}\sin\left(x + \dfrac{\pi}{6}\right) - 1$

2.5   Graphs of Other Trigonometric Functions

▶ **Graphs of the Tangent and Cotangent Functions**

The tangent and cotangent functions differ from the sine and cosine functions in significant ways:

1. Both functions have period π, so $\tan(x + \pi) = \tan x$ and $\cot(x + \pi) = \cot x$ in contrast to sine and cosine functions that have period 2π. Therefore, values of the tangent and cotangent functions repeat every π units. The patterns of their graphs repeat every π units.

2. Because $\tan x = \dfrac{\sin x}{\cos x}$, we have $\tan x = 0$ when $\sin x = 0$, and $\tan x$ is undefined when $\cos x = 0$. Now $\cot x = \dfrac{\cos x}{\sin x}$; we have $\cot x = 0$ when $\cos x = 0$ and $\cot x$ is undefined when $\sin x = 0$. Thus, $\tan x$ is undefined at $x = \pm\dfrac{\pi}{2}, \pm\dfrac{3\pi}{2}, \pm\dfrac{5\pi}{2}, \dots$, and $\cot x$ is undefined at $x = 0, \pm\pi, \pm2\pi, \dots$

3. Both functions have no amplitude; that is, there are no minimum or maximum y-values. This means both functions have the range of $(-\infty, \infty)$.

4. Both functions are odd functions, because $\tan(-x) = \dfrac{\sin(-x)}{\cos(-x)} = \dfrac{-\sin x}{\cos x} = -\tan x$, and

$\cot(-x) = \dfrac{\cos(-x)}{\sin(-x)} = \dfrac{\cos x}{-\sin x} = -\cot x$. This means that both functions are symmetric with respect to the origin.

Let's start with the basic tangent and cotangent functions. The following is the list of properties of the graphs of $y = \tan x$ and $y = \cot x$.

- For both graphs, the period is π. You can see that the curves are apart from each other by π units, and that asymptotes are also apart from each other by π units in the graphs below.

- For both graphs, the x-intercepts and asymptotes alternate on the x-axis evenly apart (every half period).

- Both graphs pass through the high key point $\left(\dfrac{\pi}{4}, 1\right)$ and low key point $\left(-\dfrac{\pi}{4}, -1\right)$.

- At $x = 0$, the tangent graph has the (mid) key point, but the cotangent graph has a key asymptote.

- At $0, \pm\pi, \pm 2\pi, \ldots$, for the tangent graph, key points repeat. On the other hand, for the cotangent graph, asymptotes repeat.

- The tangent function is increasing. On the other hand, the cotangent function is decreasing.

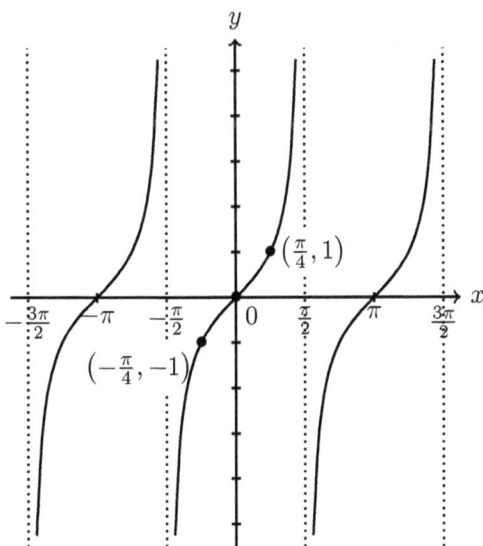

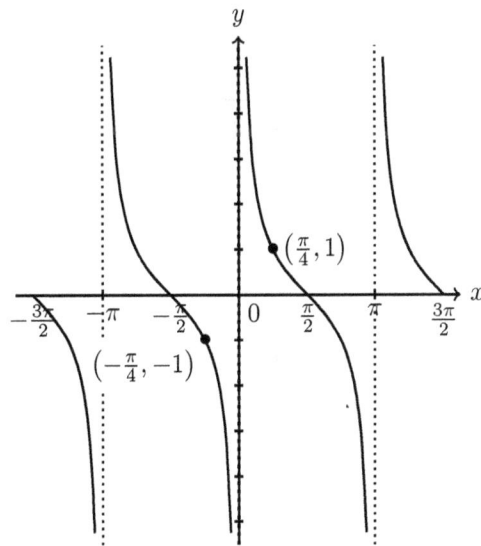

> ✏️ **Properties of Transformed Tangent and Cotangent Graphs**
>
> For $y = A \tan B(x - h) + k$ or $y = A \cot B(x - h) + k$, the graph has the following properties.
>
> - vertical stretch factor = $|A|$, (if $A < 0$, the graph is turned upside down.)
>
> - period = $\dfrac{\pi}{B}$
>
> - phase shift = h
>
> - vertical shift = k

The procedure for graphing $y = A \tan B(x - h) + k$ is based on the essential features of the graph $y = \tan x$.

Step 1. Plot the shifted origin at at (h, k). It is a **key point**. Plot key points repeatedly every period.

Step 2. Draw asymptotes between key points.

Step 3. From the key point, move **right** halfway toward the asymptote, and move up by $|A|$ (or down if $A < 0$). Plot the high (or low) key point there.

Step 4. Connect key points to draw a curve and repeat every period.

The procedure for graphing $y = A \cot B(x - h) + k$ is similar, but slightly different.

Step 1. Locate the shifted origin at (h, k). Draw an **asymptote** passing through the point and draw asymptotes repeatedly every period.

Step 2. Draw mid key points between asymptotes at the level $y = k$.

Step 3. From the mid key point, move **left** halfway toward the asymptote, and move up by $|A|$ (or down if $A < 0$). Plot the high (or low) key point there.

Step 4. Connect key points to draw a curve and repeat every period.

Example 1. Graphing Tangent and Cotangent with Different Period, Phase Shift, and Vertical Shift

Graph the function.

a) $y = 2 \tan \left(\dfrac{3}{2} x - \dfrac{\pi}{2} \right) - 1$

b) $y = 2 \cot \left(\dfrac{3}{2} x - \dfrac{\pi}{2} \right) - 1$

Solution. The equation $y = 2 \tan \left(\dfrac{3}{2} x - \dfrac{\pi}{2} \right) - 1$ can be written as $y = 2 \tan \dfrac{3}{2} \left(x - \dfrac{\pi}{3} \right) - 1$; thus, $A = 2$, $B = \dfrac{3}{2}$, $h = \dfrac{\pi}{3}$, and $k = -1$. Therefore, the vertical stretch factor is $|2| = 2$, the period is $\dfrac{\pi}{B} = \dfrac{\pi}{\frac{3}{2}} = \dfrac{2\pi}{3}$, the phase shift is $\dfrac{\pi}{3}$, and the vertical shift is -1. The same is true for the cotangent function. Follow the steps described above, beginning with the point at $(h, k) = \left(\dfrac{\pi}{3}, -1 \right)$.

a) The tangent graph is drawn as follows. Shift the origin to $\left(\dfrac{\pi}{3}, -1\right)$. Begin with a key point at the shifted origin. Draw curves and asymptotes every period. Find high and low key points using the vertical stretch factor $A = 2$.

$$y = 2\tan\left(\frac{3}{2}x - \frac{\pi}{2}\right) - 1$$

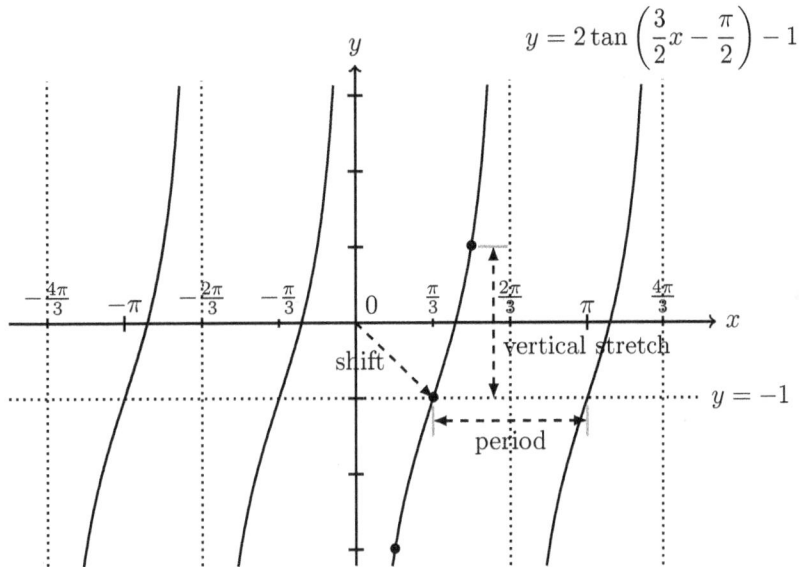

b) The cotangent graph is drawn as follows. Shift the origin to $\left(\dfrac{\pi}{3}, -1\right)$. Begin with an asymptote at the shifted origin. Draw curves and asymptotes every period. Find high and low key points using the vertical stretch factor $A = 2$.

$$y = 2\cot\left(\frac{3}{2}x - \frac{\pi}{2}\right) - 1$$

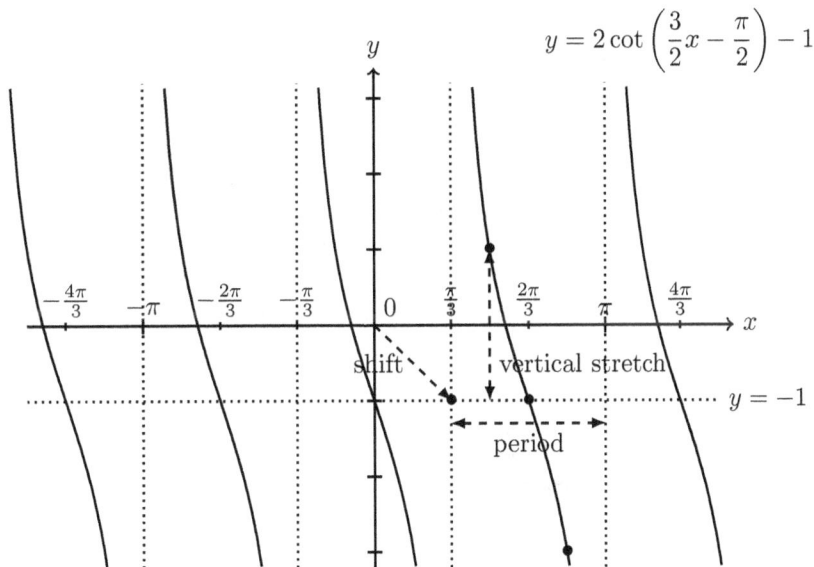

Practice Now.   Graph the function.

a) $y = 2\tan\left(\dfrac{1}{2}x - \pi\right) + 1$

b) $y = 2\cot\left(\dfrac{1}{2}x - \pi\right) + 1$

► Graphs of the Secant and Cosecant Functions

The graphs of secant and cosecant are drawn by **inverting** the graphs of sine and cosine. What inverting means is that we turn inside out the horizontal strip $-1 \le y \le 1$ (enveloping sine and cosine graphs) by taking the reciprocal of the y-coordinate, as in the picture on the right. Points in the upper half of the strip go upward and out of the strip, while points in the lower half of the strip go downward out of the strip. The closer the point is to the center, the farther the inverted point goes out. The graphs of $y = \sin x$ and $y = \cos x$ are inside the strip. Because $\sec x$ and $\csc x$ are reciprocals of $\cos x$ and $\sin x$, the graphs of $y = \sec x$ and $y = \csc x$ are drawn outside the strip.

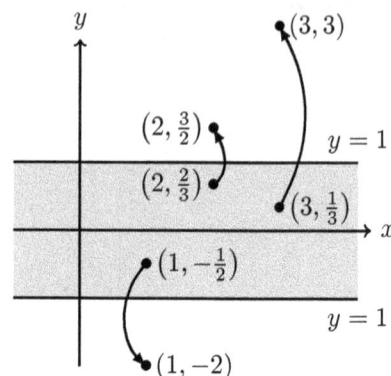

Here are some properties of the secant function in relation to the cosine function.

1. Because secant is just the reciprocal of cosine, it also has a period of 2π.

2. Because the reciprocals of 1 and -1 are themselves, secant will share the same points with cosine where $\cos x = 1$ and $\cos x = -1$.

3. Because $-1 \le \cos x \le 1$ and the reciprocals of numbers between 0 and 1 are greater than 1, either $\sec x \ge 1$ or $\sec x \le -1$.

Cosecant shares similar properties because it is the reciprocal of sine. It has a period of 2π; it shares the same points with sine where $\sin x = 1$ and $\sin x = -1$, and $\csc x \ge 1$ or $\csc x \le -1$. The graphs of the functions are as follows.

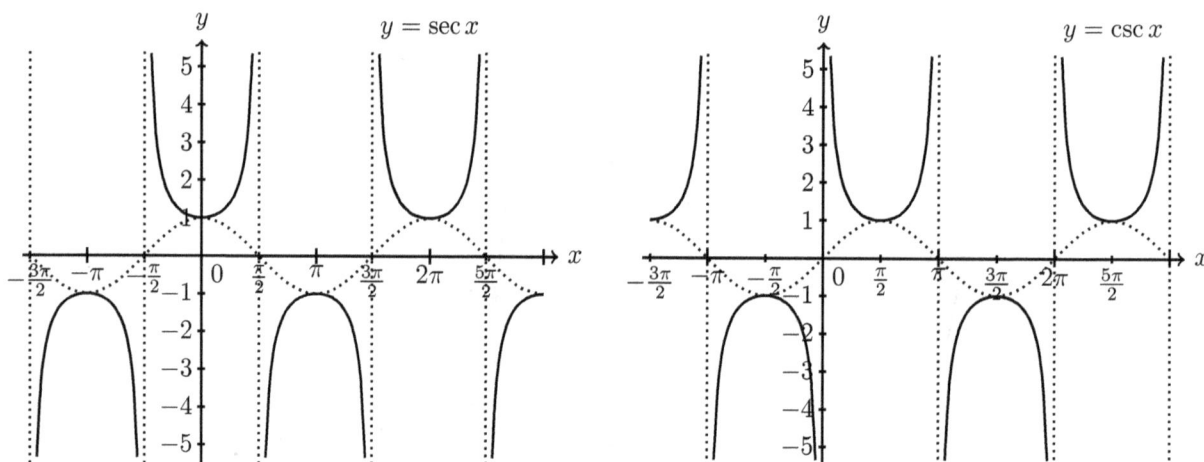

Example 2. Graphing Secant and Cosecant Functions

Graph $y = 3 \csc 2x$ over a two-period interval.

Solution.   Because cosecant is the reciprocal of sine, we graph first $y = 3 \sin 2x$ and invert the graph. Because $A = 3$ and $B = 2$, the amplitude is 3, and the period is $\dfrac{2\pi}{2} = \pi$, and there is no phase shift nor vertical shift. The graph looks as follows.

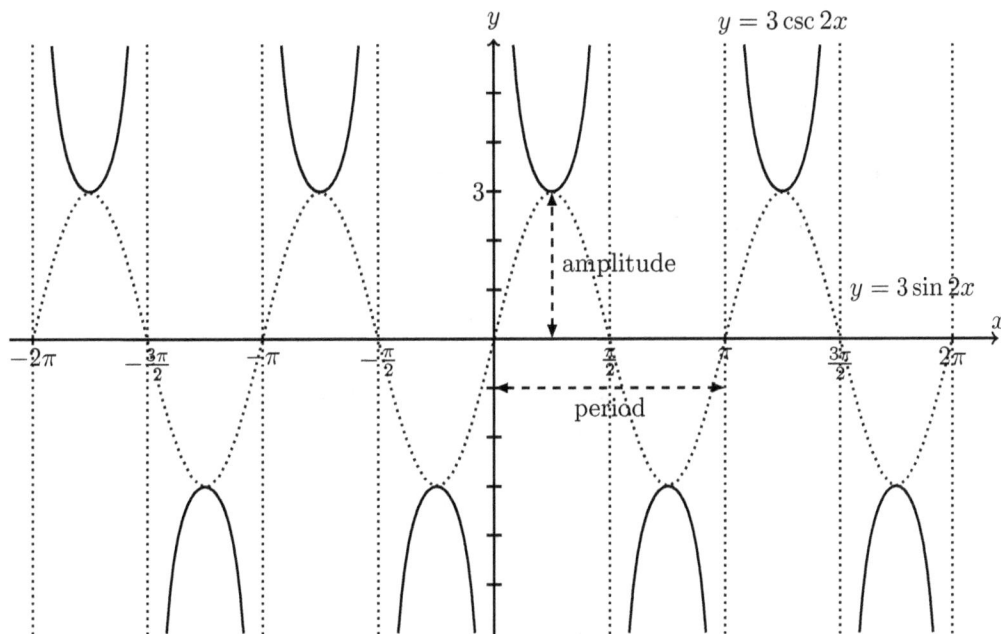

Practice Now. Graph $y = 2\sec 3x$ over a two-period interval.

Exercises 2.5

(1–11) Determine the vertical stretch factor, period, phase shift, and vertical shift of each function.

(12–21) Sketch the graph of each function.

1. $y = 2 + \cot x$

2. $y = \tan \dfrac{x}{3}$

3. $y = \cot \left(2x - \dfrac{\pi}{2}\right)$

4. $y = \dfrac{1}{3}\tan 2x$

5. $y = -5\tan(3x + \pi)$

6. $y = \tan \left(\dfrac{\pi x}{4} + 2\pi\right)$

7. $y = 2\tan \dfrac{1}{2}x - 3$

8. $y = \sec \left(x + \dfrac{\pi}{4}\right)$

9. $y = \csc(2x - \pi)$

10. $y = -\dfrac{2}{5}\tan \left(3x - \dfrac{\pi}{2}\right)$

11. $y = 3\cot \left(4x + \dfrac{\pi}{2}\right)$

12. $y = \tan(x + 2)$

13. $y = 2\cot x - 3$

14. $y = -4\tan \left(x - \dfrac{\pi}{4}\right) + 1$

15. $y = -\cot(x + \pi) - 3$

16. $y = 2\csc x + 1$

17. $y = \dfrac{1}{2}\sec 2x$

18. $y = 2\sec(3x + \pi) + 1$

19. $y = -\dfrac{2}{3}\tan 3\left(x - \dfrac{\pi}{6}\right)$

20. $y = 3\cot 4\left(x + \dfrac{\pi}{8}\right)$

21. $y = 2\tan \left(x + \dfrac{\pi}{3}\right) + 4$

2.6 Inverse Trigonometric Functions

▶ The Definitions of Inverse Trigonometric Functions

We can define the inverse function of a function only if the function is one-to-one, or, equivalently, passes the horizontal line test. But trigonometric functions do not pass the horizontal line test because they are periodic. But we can still define inverse trigonometric functions using restricted domains.

1. For $f(x) = \sin x$, we restrict the domain to $\left[-\dfrac{\pi}{2}, \dfrac{\pi}{2}\right]$. The restricted sine is a one-to-one function, and notice that it has values in the range $[-1, 1]$.

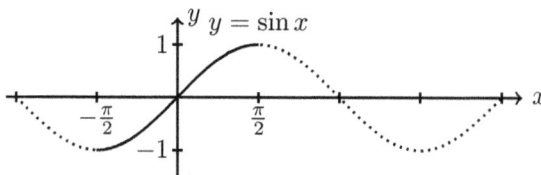

2. In a similar way, for $f(x) = \cos x$, we restrict the domain of cosine to $[0, \pi]$, so the restricted cosine is also a one-to-one function with range $[-1, 1]$.

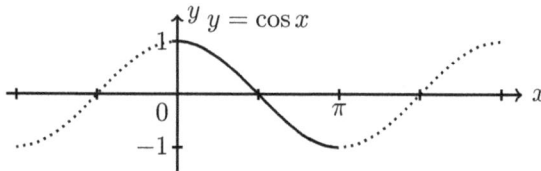

3. For $f(x) = \tan x$, we restrict the domain to $\left(-\dfrac{\pi}{2}, \dfrac{\pi}{2}\right)$, so the restricted tangent is a one-to-one function, and its range is $(-\infty, \infty)$.

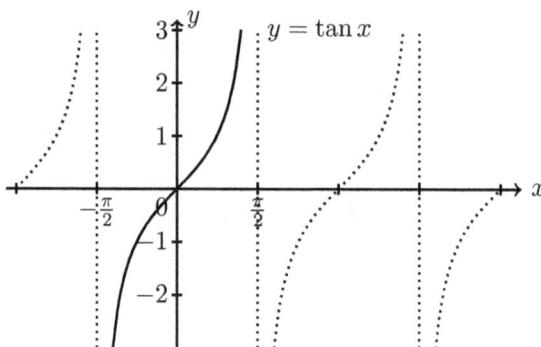

We define **inverse trigonometric functions** to be the inverse functions of the restricted trigonometric functions. We denote the inverse trigonometric functions by $\sin^{-1} x$, $\cos^{-1} x$, and $\tan^{-1} x$, or by $\arcsin x$, $\arccos x$, and $\arctan x$. Because input and output values are switched when we take inverse functions, so are domains and ranges.

> ✏ **Definitions of Inverse Trigonometric Functions**
>
> $$y = \sin^{-1} x \quad \text{if and only if} \quad \sin y = x$$
> $$y = \cos^{-1} x \quad \text{if and only if} \quad \cos y = x$$
> $$y = \tan^{-1} x \quad \text{if and only if} \quad \tan y = x$$
>
> | Function | $y = \sin^{-1} x$ | $y = \cos^{-1} x$ | $y = \tan^{-1} x$ |
> |---|---|---|---|
> | Domain (x) | $[-1, 1]$ | $[-1, 1]$ | $(-\infty, \infty)$ |
> | Range (y) | $\left[-\dfrac{\pi}{2}, \dfrac{\pi}{2}\right]$ | $[0, \pi]$ | $\left(-\dfrac{\pi}{2}, \dfrac{\pi}{2}\right)$ |

The following are the graphs of inverse trigonometric functions. They are symmetric to the graphs of trigonometric functions with respect to the line $y = x$, as is the case for any function and inverse function pair.

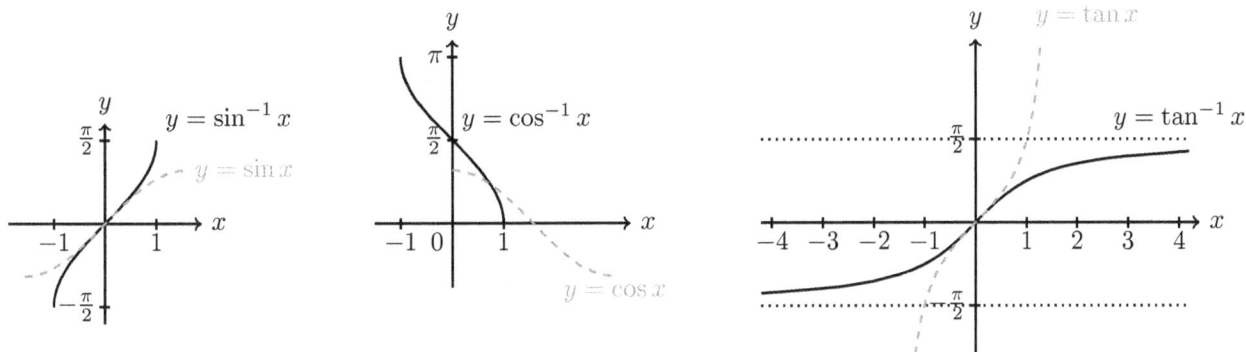

If $x > 0$, then all values of inverse sine, cosine, and tangent are some angles between 0 and $\dfrac{\pi}{2}$. But if $x < 0$, then

$$-\frac{\pi}{2} \le \sin^{-1} x < 0, \qquad \frac{\pi}{2} < \cos^{-1} x \le \pi, \qquad -\frac{\pi}{2} < \tan^{-1} x < 0.$$

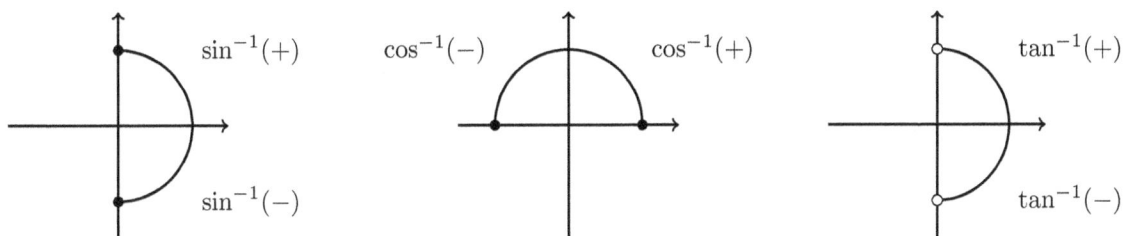

Example 1. Finding the Quadrant in which an Inverse Trigonometric Function Value Lies.

State the quadrant in which each angle lies.

a) $\sin^{-1} 0.3249$

b) $\cos^{-1}\left(-\dfrac{1}{3}\right)$

c) $\tan^{-1}(-6.32)$

Solution.

a) Because 0.3249 is positive, $\sin^{-1} 0.3249$ is an angle in quadrant I.

b) Because $-\dfrac{1}{3}$ is negative, $\cos^{-1}\left(-\dfrac{1}{3}\right)$ is an angle in quadrant II.

c) Because -6.32 is negative, $\tan^{-1}(-6.32)$ is an angle in quadrant IV.

Practice Now. State the quadrant in which each angle lies.

a) $\sin^{-1}\left(-\dfrac{3}{5}\right)$

b) $\cos^{-1} 0.3234$

c) $\tan^{-1} 0.8398$

Example 2. Finding Inverse Trigonometric Function Values by Hand

Find the exact value in radians.

a) $\sin^{-1}\left(-\dfrac{1}{2}\right)$ **b)** $\cos^{-1}\left(\dfrac{\sqrt{2}}{2}\right)$ **c)** $\cos^{-1}\left(-\dfrac{\sqrt{3}}{2}\right)$ **d)** $\tan^{-1}(-1)$

Solution.

a) Because $-\dfrac{1}{2}$ is negative, $\sin^{-1}\left(-\dfrac{1}{2}\right)$ is in the interval $\left[-\dfrac{\pi}{2},0\right)$. Remember that $\sin\dfrac{\pi}{6} = \dfrac{1}{2}$. The angle in the interval with the reference angle $\dfrac{\pi}{6}$ is $-\dfrac{\pi}{6}$. Therefore, the answer is

$$\sin^{-1}\left(-\dfrac{1}{2}\right) = -\dfrac{\pi}{6}$$

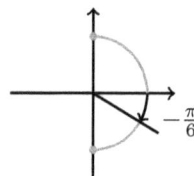

b) Because $\dfrac{\sqrt{2}}{2}$ is positive, $\cos^{-1}\left(\dfrac{\sqrt{2}}{2}\right)$ is in the interval $\left[0,\dfrac{\pi}{2}\right)$. Remember that $\cos\dfrac{\pi}{4} = \dfrac{\sqrt{2}}{2}$. The angle in the interval with the reference angle $\dfrac{\pi}{4}$ is $\dfrac{\pi}{4}$. Therefore, the answer is

$$\cos^{-1}\left(\dfrac{\sqrt{2}}{2}\right) = \dfrac{\pi}{4}$$

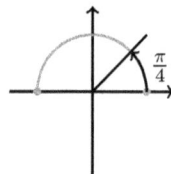

c) Because $-\dfrac{\sqrt{3}}{2}$ is negative, $\cos^{-1}\left(-\dfrac{\sqrt{3}}{2}\right)$ is in the interval $\left(\dfrac{\pi}{2},\pi\right]$. Remember that $\cos\dfrac{\pi}{6} = \dfrac{\sqrt{3}}{2}$. The angle in the interval with the reference angle $\dfrac{\pi}{6}$ is $\dfrac{5\pi}{6}$. Therefore, the answer is

$$\cos^{-1}\left(-\dfrac{\sqrt{3}}{2}\right) = \dfrac{5\pi}{6}$$

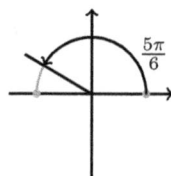

d) Because -1 is negative, $\tan^{-1}(-1)$ is in the interval $\left(-\dfrac{\pi}{2},0\right)$. Remember that $\tan\dfrac{\pi}{4} = 1$. The angle in the interval with the reference angle $\dfrac{\pi}{4}$ is $-\dfrac{\pi}{4}$. Therefore, the answer is

$$\tan^{-1}(-1) = -\dfrac{\pi}{4}$$

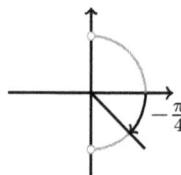

Practice Now.   Find the exact value in radians.

a) $\cos^{-1}\left(\dfrac{\sqrt{3}}{2}\right)$ **b)** $\cos^{-1}\left(-\dfrac{\sqrt{2}}{2}\right)$ **c)** $\sin^{-1}\left(\dfrac{1}{2}\right)$ **d)** $\tan^{-1}\left(-\dfrac{\sqrt{3}}{3}\right)$

Example 3. Computing Inverse Trigonometric Function Values Using the Calculator

Find the value in degrees if possible. Round to the nearest tenth.

   a) $\sin^{-1} 0.6$ **b)** $\cos^{-1}(-0.5)$ **c)** $\tan^{-1} 2$ **d)** $\sin^{-1} 2$

Solution. Be sure to put the calculator in degree mode before you type in the expression. For most calculators, you can type inverse trigonometric functions by pressing the "2nd" or shift button followed by a normal trigonometric function button.

a) If the calculator accepts and displays the whole expression in one line, type the function first: $\boxed{\text{2nd}}$, $\boxed{\sin}$, $\boxed{0.6}$, $\boxed{=}$, or for calculators of simpler type, input the value first: $\boxed{0.6}$, $\boxed{\text{2nd}}$, $\boxed{\sin}$ to get the answer $\sin^{-1} 0.6 = 36.9°$. Solve other problems similarly.

b) $\cos^{-1}(-0.5) = 120°$.

c) $\tan^{-1} 2 = 63.4°$.

d) The expression gives a domain error. This is because 2 is not in the domain of the inverse sine function, which is $[-1, 1]$. So, the expression is undefined.

Practice Now. Find the value in degrees if possible. Round to the nearest tenth.

   a) $\sin^{-1}(-0.2)$ **b)** $\cos^{-1}\left(\dfrac{2}{3}\right)$ **c)** $\tan^{-1}(-0.5432)$ **d)** $\cos^{-1}(-3)$

▶ **Composition of Trigonometric and Inverse Trigonometric Functions**

In general, when we compose a function and its inverse, they cancel each other,

$$f(f^{-1}(x)) = x, \qquad f^{-1}(f(x)) = x,$$

on one condition: x should be in the domain. For trigonometric functions, we have the following rules.

Composition of a Trigonometric Function and Its Inverse

$$\sin\left(\sin^{-1} x\right) = x \quad \text{for } -1 \le x \le 1 \qquad \sin^{-1}\left(\sin x\right) = x \quad \text{for } -\frac{\pi}{2} \le x \le \frac{\pi}{2}$$

$$\cos\left(\cos^{-1} x\right) = x \quad \text{for } -1 \le x \le 1 \qquad \cos^{-1}\left(\cos x\right) = x \quad \text{for } 0 \le x \le \pi$$

$$\tan\left(\tan^{-1} x\right) = x \quad \text{for } -\infty < x < \infty \qquad \tan^{-1}\left(\tan x\right) = x \quad \text{for } -\frac{\pi}{2} < x < \frac{\pi}{2}$$

Example 4. Evaluating Composition of a Trigonometric Function and its Inverse

Find the value of the composition.

   a) $\sin\left(\sin^{-1} 0.7382\right)$ **b)** $\cos\left(\cos^{-1} 3\right)$ **c)** $\sin^{-1}\left(\sin\dfrac{\pi}{3}\right)$ **d)** $\tan^{-1}\left(\tan\dfrac{3\pi}{4}\right)$

Solution.   We can apply the rules if x is in the domain, but we can't if it is not.

a) Check the domain. Because 0.7382 is between -1 and 1, we can apply the rule. The answer is $\sin\left(\sin^{-1} 0.7382\right) = 0.7382$.

b) Check the domain. Because 3 is not between -1 and 1, we can't apply the rule. Compute the expression inside parentheses first. Because $\cos^{-1} 3$ is undefined, so is $\cos\left(\cos^{-1} 3\right)$.

c) Check the domain. Because $\dfrac{\pi}{3}$ is between $-\dfrac{\pi}{2}$ and $\dfrac{\pi}{2}$, we can apply the rule. The answer is
$$\sin^{-1}\left(\sin\frac{\pi}{3}\right) = \frac{\pi}{3}.$$

d) Check the domain. Because $\dfrac{3\pi}{4}$ is not between $-\dfrac{\pi}{2}$ and $\dfrac{\pi}{2}$, we can't apply the rule. Compute the expression inside parentheses first. Because $\dfrac{3\pi}{4}$ is in quadrant II, where tangent values are negative, and the reference angle of $\dfrac{3\pi}{4}$ is $\dfrac{\pi}{4}$,
$$\tan\frac{3\pi}{4} = -\tan\frac{\pi}{4} = -1.$$
Then use it to get the inverse tangent value as in previous examples.
$$\tan^{-1}\left(\tan\frac{3\pi}{4}\right) = \tan^{-1}(-1) = -\frac{\pi}{4}.$$

Practice Now.   Find the value of the composition.

a) $\tan\left[\tan^{-1}(-3.428)\right]$ **b)** $\cos^{-1}\left(\cos\dfrac{2\pi}{3}\right)$ **c)** $\cos^{-1}\left[\cos\left(-\dfrac{\pi}{4}\right)\right]$

Composing a trigonometric function and an inverse trigonometric function of different type is not straightforward. But we can still get the exact value of the composition by sketching the angle.

Example 5. Composing Trigonometric and Inverse Trigonometric Functions of Different Type

Find the exact value.

a) $\cos\left[\sin^{-1}\left(-\dfrac{3}{5}\right)\right]$ **b)** $\sin\left(\tan^{-1} 2\right)$ **c)** $\sin^{-1}\left(\cos\dfrac{\pi}{6}\right)$

Solution.

a) Let $\theta = \sin^{-1}\left(-\dfrac{3}{5}\right)$. Then, by definition, $\sin\theta = -\dfrac{3}{5}$. We are looking for $\cos\theta$. Because θ is an angle in quadrant IV ($-3/5$ is negative), sketch the angle as follows.

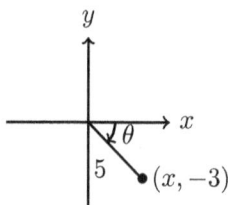

By the definition of the sine function $\sin\theta = \dfrac{y}{r}$, we may assume $y = -3$ and $r = 5$, and solve the equation $x^2 + y^2 = r^2$ for the value of x.
$$x^2 + (-3)^2 = 5^2 \quad\Longrightarrow\quad x = \pm 4$$
Because the angle is in quadrant IV, $x > 0$. Therefore, $x = 4$.

Then the answer is $\cos\left[\sin^{-1}\left(-\dfrac{3}{5}\right)\right] = \cos\theta = \dfrac{x}{r} = \dfrac{4}{5}$.

b) Let $\theta = \tan^{-1} 2$. Then, by definition, $\tan \theta = 2$. We are looking for $\sin \theta$. Because θ is an angle in quadrant I (2 is positive), sketch the angle as follows.

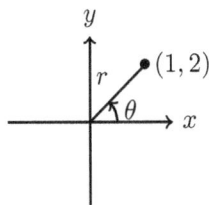

By the definition of the tangent function $\tan \theta = \dfrac{y}{x}$, we may assume $y = 2$ and $x = 1$, and use the formula for r.

$$r = \sqrt{x^2 + y^2} = \sqrt{1^2 + 2^2} = \sqrt{5}.$$

Then the answer is $\sin \left(\tan^{-1} 2 \right) = \sin \theta = \dfrac{y}{r} = \dfrac{2}{\sqrt{5}} = \dfrac{2\sqrt{5}}{5}.$

c) We are looking for the angle in the interval $[0, \pi]$ whose sine is the same as the $\cos \dfrac{\pi}{6}$. We know that $\cos \dfrac{\pi}{6} = \dfrac{\sqrt{3}}{2}$. We also know that $\dfrac{\pi}{3}$ is the angle in $[0, \pi]$ such that $\sin \dfrac{\pi}{3} = \dfrac{\sqrt{3}}{2}$. Therefore,

$$\sin^{-1} \left(\cos \dfrac{\pi}{6} \right) = \sin^{-1} \left(\dfrac{\sqrt{3}}{2} \right) = \dfrac{\pi}{3}.$$

Practice Now. Find the exact value, without using a calculator.

a) $\cos \left[\sin^{-1} \left(\dfrac{\sqrt{3}}{2} \right) \right]$ **b)** $\tan \left[\cos^{-1} \left(-\dfrac{\sqrt{2}}{2} \right) \right]$ **c)** $\sin^{-1} \left(\cos \dfrac{\pi}{4} \right)$

Exercises 2.6

(1–9) Evaluate the given expression without the aid of a calculator.

1. $\sin^{-1} \left(\dfrac{1}{2} \right)$

2. $\cos^{-1} \left(-\dfrac{\sqrt{3}}{2} \right)$

3. $\tan^{-1} \left(-\sqrt{3} \right)$

4. $\sin^{-1} \left(-\dfrac{\sqrt{2}}{2} \right)$

5. $\cos^{-1} \left(-\dfrac{1}{2} \right)$

6. $\tan^{-1} 1$

7. $\cot^{-1} \sqrt{3}$

8. $\sin^{-1} 1 - \cos^{-1} \left(-\dfrac{1}{2} \right)$

9. $\sin^{-1} \left(\dfrac{1}{2} \right) + \cos^{-1} 1$

(10–23) Find the exact value without a calculator.

10. $\sin \left[\cos^{-1} \left(\dfrac{12}{13} \right) \right]$

11. $\sin \left[\tan^{-1} \left(\dfrac{3}{4} \right) \right]$

12. $\cos \left[\tan^{-1} \left(-\dfrac{4}{3} \right) \right]$

13. $\cos^{-1} \left[\sin \left(\dfrac{\pi}{6} \right) \right]$

14. $\cos \left[\sin^{-1} \left(-\dfrac{1}{2} \right) \right]$

15. $\sin^{-1} \left[\sin \left(\dfrac{\pi}{3} \right) \right]$

16. $\sin \left(\tan^{-1} \sqrt{3} \right)$

17. $\cos^{-1} \left[2 \sin \left(\dfrac{\pi}{6} \right) \right]$

18. $\cos \left[2 \sin^{-1} \left(\dfrac{1}{2} \right) \right]$

19. $\sin \left[2 \tan^{-1} \left(\dfrac{5}{12} \right) \right]$

20. $\sin^{-1} \left[\cos \left(-\dfrac{\pi}{3} \right) \right]$

21. $\cos \left[\tan^{-1} \left(-1 \right) \right]$

22. $\tan \left[\cos^{-1} \left(-\dfrac{1}{2} \right) \right]$

23. $\cos \left[\sin^{-1} \left(\dfrac{3}{5} \right) \right]$

(24–28) Use the calculator to find the following angles in degrees if they exist.

24. $\sin^{-1} 0.6072$

25. $\sin^{-1}(-0.3276)$

26. $\cos^{-1}(-0.4781)$

27. $\tan^{-1}(-1.1761)$

28. $\cos^{-1} 2.7681$

Chapter 3

TRIGONOMETRIC IDENTITIES AND EQUATIONS

3.1 Verifying Identities

An **identity** is a statement in which two expressions are equal for every value of the variable; for example, $x + x = 2x$ is an identity. The left-hand expression always equals the right-hand expression no matter what x equals.

In this section, we learn how to use fundamental identities to simplify trigonometric expressions or to verify more complicated trigonometric identities. The following is the list of fundamental identities. They can be derived directly from the definitions of trigonometric functions.

The Fundamental Identities

- Reciprocal Identities

$$\sin \theta = \frac{1}{\csc \theta} \qquad \cos \theta = \frac{1}{\sec \theta} \qquad \tan \theta = \frac{1}{\cot \theta}$$

$$\csc \theta = \frac{1}{\sin \theta} \qquad \sec \theta = \frac{1}{\cos \theta} \qquad \cot \theta = \frac{1}{\tan \theta}$$

- Quotient Identities

$$\tan \theta = \frac{\sin \theta}{\cos \theta} \qquad \cot \theta = \frac{\cos \theta}{\sin \theta}$$

- Negative Angle Identities

$$\sin(-x) = -\sin x \qquad \cos(-x) = \cos x \qquad \tan(-x) = -\tan x$$

$$\csc(-x) = -\csc x \qquad \sec(-x) = \sec x \qquad \cot(-x) = -\cot x$$

- Pythagorean Identities

$$\sin^2 x + \cos^2 x = 1 \quad \Rightarrow \quad 1 - \sin^2 x = \cos^2 x \quad \text{or} \quad 1 - \cos^2 x = \sin^2 x$$

$$\tan^2 x + 1 = \sec^2 x \quad \Rightarrow \quad \tan^2 x = \sec^2 x - 1 \quad \text{or} \quad \sec^2 x - \tan^2 x = 1$$

$$1 + \cot^2 x = \csc^2 x \quad \Rightarrow \quad \cot^2 x = \csc^2 x - 1 \quad \text{or} \quad \csc^2 x - \cot^2 x = 1$$

A few helpful tips on verifying identities follow.

- We are not "solving" identities. We are "explaining why" they are identities. So, we must approach identities differently.

- Work with one side at a time. Do not carry both sides around.

- Start with simplifying the more complicated side.

- We want both sides to be exactly the same in the end.

- Use algebraic manipulations and/or the fundamental identities until you have the same expression as on the other side.

- Try using reciprocal/quotient identities to change all functions to be in terms of sine and cosine.

Example 1. Simplifying Trigonometric Expressions

Use the fundamental identities to simplify the expression.

a) $\dfrac{\sin x}{\csc x} + \dfrac{\cos x}{\sec x}$

b) $(1 - \sin x)(1 + \csc x)$

Solution.

a) In each step, apply one of the fundamental identities or apply algebra rules.

$$\dfrac{\sin x}{\csc x} + \dfrac{\cos x}{\sec x} = \dfrac{\sin x}{\frac{1}{\sin x}} + \dfrac{\cos x}{\frac{1}{\cos x}} \qquad \text{(Reciprocal identities)}$$

$$= \sin x \cdot \dfrac{\sin x}{1} + \cos x \cdot \dfrac{\cos x}{1} \qquad \text{(Simplify double fractions by multiplying reciprocals.)}$$

$$= \sin^2 x + \cos^2 x \qquad \text{(Simplify.)}$$

$$= 1 \qquad \text{(Pythagorean identity)}$$

b) The same idea applies to this problem.

$$(1 - \sin x)(1 + \csc x) = 1 + \csc x - \sin x - \sin x \csc x \qquad \text{(Expand.)}$$

$$= 1 + \csc x - \sin x - \sin x \cdot \dfrac{1}{\sin x} \qquad \text{(Reciprocal identity)}$$

$$= 1 + \csc x - \sin x - 1 \qquad \text{(Simplify.)}$$

$$= \csc x - \sin x \qquad \text{(Simplify.)}$$

Practice Now.   Use the fundamental identities to simplify the expression.

a) $\sin(-x)\sin x - \cos(-x)\cos x$

b) $(1 - \sec x)(1 + \sec x)$

Example 2. Verifying a Trigonometric Identity

Verify each trigonometric identity.

a) $\cot x \sin x = \cos x$ **b)** $\dfrac{\cos x}{1 - \sin x} + \dfrac{\cos x}{1 + \sin x} = \dfrac{2\sec^3 x}{1 + \tan^2 x}$ **c)** $\dfrac{\sin x}{1 + \cos x} = \csc x - \cot x$

Solution.

a) It is better to start simplifying the left-hand side (LHS), which looks more complicated than the right-hand side (RHS).

$$\text{LHS} = \cot x \sin x$$
$$= \frac{\cos x}{\sin x} \cdot \sin x \qquad \text{(Quotient identity)}$$
$$= \cos x \qquad \text{(Cancel the common factor.)}$$

The simplified expression is the same as the expression on the right-hand side. The identity is verified.

b) Simplify both sides one by one, and try to get the same result at the end.

$$\text{LHS} = \frac{\cos x}{1 - \sin x} + \frac{\cos x}{1 + \sin x}$$
$$= \frac{\cos x}{1 - \sin x} \cdot \frac{1 + \sin x}{1 + \sin x} + \frac{\cos x}{1 + \sin x} \cdot \frac{1 - \sin x}{1 - \sin x} \qquad \text{(Make a common denominator.)}$$
$$= \frac{(\cos x - \cos x \sin x) + (\cos x + \cos x \sin x)}{1 - \sin^2 x} \qquad \text{(Add fractions.)}$$
$$= \frac{2 \cos x}{1 - \sin^2 x} \qquad \text{(Simplify the numerator.)}$$
$$= \frac{2 \cos x}{\cos^2 x} \qquad \text{(Pythagorean identity)}$$
$$= \frac{2}{\cos x} \qquad \text{(Cancel the common factor.)}$$
$$= 2 \sec x \qquad \text{(Reciprocal identity)}$$

$$\text{RHS} = \frac{2 \sec^3 x}{1 + \tan^2 x}$$
$$= \frac{2 \sec^3 x}{\sec^2 x} \qquad \text{(Pythagorean identity)}$$
$$= 2 \sec x \qquad \text{(Cancel the common factor.)}$$

Because both sides are simplified to be the same, the identity is verified.

c) At first glance, the left-hand side is already an expression of sine and cosine, and it is not clear what to do with it. So, start with the right-hand side.

$$\text{RHS} = \csc x - \cot x$$
$$= \frac{1}{\sin x} - \frac{\cos x}{\sin x} \qquad \text{(Reciprocal and quotient identities)}$$
$$= \frac{1 - \cos x}{\sin x} \qquad \text{(Add fractions.)}$$

At this point, comparing the result and the left-hand side reveals that they do not look exactly the same. We need a new trick to verify this identity. Sometimes using the conjugate helps in verifying an identity. Make a conjugate by changing the sign between two terms. The conjugate of $1 - \cos x$ is $1 + \cos x$, and vice versa. So, continue working the right-hand side by multiplying the conjugate as follows.

$$= \frac{1 - \cos x}{\sin x} \cdot \frac{1 + \cos x}{1 + \cos x} \qquad \text{(Multiply by conjugates.)}$$

$$= \frac{1 - \cos^2 x}{\sin x(1 + \cos x)} \qquad \text{(Simplify.)}$$

$$= \frac{\sin^2 x}{\sin x(1 + \cos x)} \qquad \text{(Pythagorean identity)}$$

$$= \frac{\sin x}{1 + \cos x} \qquad \text{(Cancel the common factor.)}$$

Now it looks exactly like the left-hand side. The identity is verified.

Practice Now.   Establish the identities.

a) $\sin x \csc x - \cos^2 x = \sin^2 x$

b) $\dfrac{\cos x + 1}{\cos x - 1} = \dfrac{1 + \sec x}{1 - \sec x}$

c) $\dfrac{1 + \sin x}{1 - \sin x} - \dfrac{1 - \sin x}{1 + \sin x} = 4\tan x \sec x$

d) $\cot x + \tan x = \dfrac{\cos x}{\sin x} + \dfrac{\sin x}{\cos x}$

Exercises 3.1

(1–14) Simplify each trigonometric expression.

1. $\sin x \csc x$

2. $\cot x \tan x$

3. $\sec x \tan x \cos x$

4. $\sin^2 x \cot x \csc x$

5. $\dfrac{1 - \sin^2 x}{\cos^2 x}$

6. $\dfrac{1 + \cot^2 x}{\csc^2 x}$

7. $\dfrac{(1 - \sec x)(1 + \sec x)}{\tan x}$

8. $(1 + \cot x)^2 - 2\cot x$

9. $\dfrac{1 - \tan^2 x}{1 + \tan x}$

10. $\dfrac{1 - \cos^2 x}{1 - \cos x}$

11. $\dfrac{1 + \sin x \sin(-x)}{\cos x}$

12. $\dfrac{\tan x}{1 + \sec x \sec(-x)}$

13. $\dfrac{1 + \tan x}{\sec x}$

14. $1 + \dfrac{\csc x}{\sin x}$

(15–37) Verify the following identities.

15. $\cot x + \tan x = \csc x \sec x$

16. $\tan x \sin x + \cos x = \sec x$

17. $\dfrac{\csc^2 x - 1}{\csc^2 x} = \cos^2 x$

18. $\dfrac{1}{\tan x} + \tan x = \dfrac{1}{\sin x \cos x}$

19. $\sin x - \sin x \cos^2 x = \sin^3 x$

20. $\tan x + \dfrac{\cos x}{1 + \sin x} = \dfrac{1}{\cos x}$

21. $\dfrac{\cos x}{1 + \sin x} + \dfrac{1 + \sin x}{\cos x} = 2\sec x$

22. $1 - 2\cos^2 x = \dfrac{\tan^2 x - 1}{\tan^2 x + 1}$

23. $\tan^2 x = \csc^2 x \tan^2 x - 1$

24. $\dfrac{\sec x - 1}{\sec x + 1} = \dfrac{1 - \cos x}{1 + \cos x}$

25. $\dfrac{1}{1 - \sin x} - \dfrac{1}{1 + \sin x} = 2\tan x \sec x$

26. $\dfrac{1 + \tan^2 x}{1 - \tan^2 x} = \dfrac{1}{\cos^2 x - \sin^2 x}$

27. $\sin^4 x - \cos^4 x = 2\sin^2 x - 1$

28. $\dfrac{\csc x}{\sin x} - \dfrac{\cot x}{\tan x} = 1$

29. $(\sin x - \cos x)^2 + (\sin x + \cos x)^2 = 2$

30. $\dfrac{\cos x}{1 - \sin x} - \tan x = \sec x$

31. $\cos^2 x = \dfrac{\csc x \cos x}{\tan x + \cot x}$

32. $\sec x + \tan x = \dfrac{\cos x}{1 - \sin x}$

33. $\sin x(1 + \csc x) = \sin x + 1$

34. $\cos^2 x - \sin^2 x = 2\cos^2 x - 1$

35. $(\sec x + \tan x)(\sec x - \tan x) = 1$

36. $\dfrac{1 + \tan x}{1 - \tan x} = \dfrac{\cos x + \sin x}{\cos x - \sin x}$

37. $(\sin x - \tan x)(\cos x - \cot x) = (\sin x - 1)(\cos x - 1)$

3.2   Sum and Difference Identities

The sum and difference identities are important in calculus and useful in certain applications. As a first example, we will find exact trigonometric values of certain angles. Although a calculator can be used to find an approximation for $\cos 15°$, for example, the identities shown below can be applied to get the exact value.

Sum and Difference Identities

Sine of a Sum or Difference

$$\sin(u + v) = \sin u \cos v + \cos u \sin v$$
$$\sin(u - v) = \sin u \cos v - \cos u \sin v$$

Cosine of a Sum or Difference

$$\cos(u + v) = \cos u \cos v - \sin u \sin v$$
$$\cos(u - v) = \cos u \cos v + \sin u \sin v$$

Tangent of a Sum or Difference

$$\tan(u + v) = \frac{\tan u + \tan v}{1 - \tan u \tan v}$$
$$\tan(u - v) = \frac{\tan u - \tan v}{1 + \tan u \tan v}$$

Example 1. Using Sum and Difference Identities to Find Exact Values of Trigonometric Functions

Find the exact value of each expression.

a) $\cos 15°$ **b)** $\sin \dfrac{5\pi}{12}$ **c)** $\tan \dfrac{7\pi}{12}$

Solution. The first step is to express the angle as a sum or difference of familiar angles.

a) To find $\cos 15°$, we write $15°$ as the sum or difference of two angles with known function values. Because we know the exact trigonometric function values of $45°$ and $30°$, we write $15°$ as $45° - 30°$. Then,

$$
\begin{aligned}
\cos 15° &= \cos\left(45° - 30°\right) \\
&= \cos 45° \cos 30° + \sin 45° \sin 30° \qquad \text{(Difference identity for cosine)} \\
&= \frac{\sqrt{2}}{2} \cdot \frac{\sqrt{3}}{2} + \frac{\sqrt{2}}{2} \cdot \frac{1}{2} \\
&= \frac{\sqrt{6} + \sqrt{2}}{4}
\end{aligned}
$$

b) We write $\dfrac{5\pi}{12}$ as the sum of $\dfrac{\pi}{4}$ and $\dfrac{\pi}{6}$ $\left(\text{in degree measurement } \dfrac{5\pi}{12} = 75° = 45° + 30° = \dfrac{\pi}{4} + \dfrac{\pi}{6}\right)$. So

$$
\begin{aligned}
\sin \frac{5\pi}{12} &= \sin\left(\frac{\pi}{4} + \frac{\pi}{6}\right) \\
&= \sin \frac{\pi}{4} \cos \frac{\pi}{6} + \cos \frac{\pi}{4} \sin \frac{\pi}{6} \qquad \text{(Sum identity for sine)} \\
&= \frac{\sqrt{2}}{2} \cdot \frac{\sqrt{3}}{2} + \frac{\sqrt{2}}{2} \cdot \frac{1}{2} \\
&= \frac{\sqrt{6} + \sqrt{2}}{4}
\end{aligned}
$$

c) We write $\dfrac{\pi}{12}$ as the difference of $\dfrac{\pi}{3}$ and $\dfrac{\pi}{4}$ $\left(\text{in degree measurement } \dfrac{\pi}{12} = 15° = 60° - 45° = \dfrac{\pi}{3} - \dfrac{\pi}{4}\right)$. So

$$
\begin{aligned}
\tan \frac{\pi}{12} &= \tan\left(\frac{\pi}{3} - \frac{\pi}{4}\right) \\
&= \frac{\tan\dfrac{\pi}{3} - \tan\dfrac{\pi}{4}}{1 + \tan\dfrac{\pi}{3}\tan\dfrac{\pi}{4}} \qquad \text{(Difference identity for tangent)} \\
&= \frac{\sqrt{3} - 1}{1 + \sqrt{3}\cdot 1} \\
&= \frac{\sqrt{3} - 1}{1 + \sqrt{3}}\cdot\frac{1 - \sqrt{3}}{1 - \sqrt{3}} \qquad \text{(Rationalize the denominator.)} \\
&= \frac{\sqrt{3} - 3 - 1 + \sqrt{3}}{1 - 3} \\
&= \frac{2\sqrt{3} - 4}{-2} \\
&= \frac{2(\sqrt{3} - 2)}{-2} \\
&= -\sqrt{3} + 2
\end{aligned}
$$

Practice Now.   Find the exact value of each expression.

a) $\cos\dfrac{5\pi}{12}$ 　　　　　　　　　 **b)** $\sin 75°$ 　　　　　　　　　 **c)** $\tan 105°$

Example 2. Using Sum and Difference identities to Simplify Expressions

Find the exact value of each expression.

a) $\cos 87° \cos 93° - \sin 87° \sin 93°$

b) $\sin 40° \cos 160° - \cos 40° \sin 160°$

c) $\dfrac{\tan 80° - \tan(-55°)}{1 + \tan 80° \tan(-55°)}$

Solution.   Trigonometric identities are used both forward and backward. Use them backward for these problems.

a) Use the sum identity to simplify and evaluate the expression.

$$
\begin{aligned}
\cos 87° \cos 93° - \sin 87° \sin 93° &= \cos(87° + 93°) \qquad \text{(Sum identity for cosine)} \\
&= \cos 180° \\
&= -1
\end{aligned}
$$

b) Use the difference identity to simplify and evaluate the expression.

$$
\begin{aligned}
\sin 40° \cos 160° - \cos 40° \sin 160° &= \sin(40° - 160°) \qquad \text{(Difference identity for sine)} \\
&= \sin(-120°) \\
&= -\sin 60° \qquad \text{(Quadrant III and reference angle 60°)} \\
&= -\frac{\sqrt{3}}{2}
\end{aligned}
$$

c) Use the difference identity to simplify and evaluate the expression.

$$\frac{\tan 80° - \tan(-55°)}{1 + \tan 80° \tan(-55°)} = \tan(80° - (-55°)) \qquad \text{(Difference identity for tangent)}$$

$$= \tan 135°$$

$$= -\tan 45° \qquad \text{(Quadrant II and reference angle 45°)}$$

$$= -1$$

Practice Now.   Find the exact value of each expression.

a) $\cos 76° \cos 31° - \sin 76° \sin 31°$

b) $\sin 63° \cos 27° - \cos 63° \sin 27°$

c) $\dfrac{\tan 75° - \tan 30°}{1 + \tan 75° \tan 30°}$

Another use of sum and difference identities is to prove a set of cofunction identities.

> ### ✎ Cofunction Identities
>
> The trigonometric function value of the complement of an angle is equal to the cofunction value of the angle.
>
> $$\sin(90° - \theta) = \cos\theta \qquad\qquad \cos(90° - \theta) = \sin\theta$$
> $$\tan(90° - \theta) = \cot\theta \qquad\qquad \cot(90° - \theta) = \tan\theta$$
> $$\sec(90° - \theta) = \csc\theta \qquad\qquad \csc(90° - \theta) = \sec\theta$$

Example 3. Using Sum and Difference Identities to Verify Cofunction Identities

Verify the identity.

a) $\sin(90° - \theta) = \cos\theta$

b) $\cos(\pi + \theta) = -\cos\theta$

Solution.   Apply sum and difference identities.

a) Use the sine difference identity.

$$\sin(90° - \theta) = \sin 90° \cos\theta - \cos 90° \sin\theta$$
$$= (1)\cos\theta - (0)\sin\theta$$
$$= \cos\theta$$

b) Use the cosine sum identity.

$$\cos(\pi + \theta) = \cos\pi \cos\theta - \sin\pi \sin\theta$$
$$= (-1)\cos\theta - (0)\sin\theta$$
$$= -\cos\theta$$

Practice Now.   Verify the identity.

a) $\cos\left(180° - \theta\right) = -\cos\theta$

b) $\sin\left(\dfrac{3\pi}{2} + \theta\right) = -\cos\theta$

Example 4. Using Cofunction Identities to Find an Angle

Find an angle θ that satisfies the equation.

a) $\cot\theta = \tan 25°$

b) $\sin\theta = \cos\left(-30°\right)$

c) $\csc\dfrac{3\pi}{4} = \sec\theta$

Solution.

a) Because tangent and cotangent are cofunctions, we have $\tan\left(90° - \theta\right) = \cot\theta$.

$$\cot\theta = \tan 25°$$
$$\tan\left(90° - \theta\right) = \tan 25° \qquad\qquad \text{(Cofunction identity)}$$
$$90° - \theta = 25°$$
$$\theta = 65°$$

b) Because sine and cosine are cofunctions, we have $\cos\left(90° - \theta\right) = \sin\theta$.

$$\sin\theta = \cos\left(-30°\right)$$
$$\cos\left(90° - \theta\right) = \cos\left(-30°\right) \qquad\qquad \text{(Cofunction identity)}$$
$$90° - \theta = -30°$$
$$\theta = 120°$$

c) Because secant and cosecant are cofunctions, we have $\sec\left(\dfrac{\pi}{2} - \theta\right) = \csc\theta$.

$$\csc\dfrac{3\pi}{4} = \sec\theta$$
$$\sec\left(\dfrac{\pi}{2} - \dfrac{3\pi}{4}\right) = \sec\theta \qquad\qquad \text{(Cofunction identity)}$$
$$\sec\left(-\dfrac{\pi}{4}\right) = \sec\theta$$
$$\theta = -\dfrac{\pi}{4}$$

Practice Now.   Find an angle θ that satisfies the equation.

a) $\tan\theta = \cot 35°$

b) $\cos\theta = \sin 75°$

c) $\sec\dfrac{\pi}{4} = \csc\theta$

Example 5. Using Sum and Difference Identities with the Values Given Partially

Given $\sin u = \dfrac{4}{5}$ and $\cos v = \dfrac{5}{13}$ with $\dfrac{\pi}{2} \leq u \leq \pi$ and $\dfrac{3\pi}{2} \leq v \leq 2\pi$, find each trigonometric function value.

a) $\cos\left(u - v\right)$

b) $\sin\left(u + v\right)$

c) $\tan\left(u - v\right)$

Solution.   The sum and difference identities require the values of $\sin u$, $\cos u$, $\tan u$, $\sin v$, $\cos v$, and $\tan v$. We are given the values of only $\sin u$ and $\cos v$. So, we must find other values first. There are two methods to do this—one is using fundamental identities and the other is sketching an angle. Both are demonstrated below. But you may choose a method you prefer.

- Using fundamental identities
  To find $\cos u$ from the given value of $\sin u$, use the Pythagorean identity $\sin^2 u + \cos^2 u = 1$.

$$\left(\frac{4}{5}\right)^2 + \cos^2 u = 1$$

$$\frac{16}{25} + \cos^2 u = 1$$

$$\cos^2 u = 1 - \frac{16}{25} = \frac{9}{25}$$

$$\cos u = \pm\sqrt{\frac{9}{25}} = \pm\frac{3}{5}$$

Because $\frac{\pi}{2} \le u \le \pi$, the angle u lies in quadrant II, where cosine is negative. Therefore, we take the negative solution, $\cos u = -\frac{3}{5}$. Next, to find $\tan u$, use the quotient identity.

$$\tan u = \frac{\sin u}{\cos u} = \frac{\dfrac{4}{5}}{-\dfrac{3}{5}} = \frac{4}{5} \cdot \left(-\frac{5}{3}\right) = -\frac{4}{3}$$

- Using a sketch

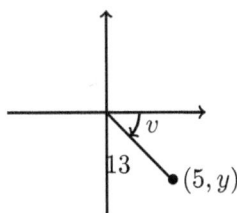

  Draw the angle v in quadrant IV because $\frac{3\pi}{2} \le v \le 2\pi$. By the definition of the cosine function $\cos v = \frac{x}{r}$, we may assume $x = 5$ and $r = 13$, and solve the equation $x^2 + y^2 = r^2$ for the value of y.

$$(5)^2 + y^2 = 13^2 \quad \Longrightarrow \quad y = \pm 12$$

  Because the angle is in quadrant IV, $y < 0$. Therefore, $y = -12$.
  Now use the sketch and the definitions of trigonometric functions to get

$$\sin v = -\frac{12}{13} \quad \text{and} \quad \tan v = -\frac{12}{5}.$$

a) Use the formula $\cos(u - v) = \cos u \cos v + \sin u \sin v$.

$$\cos(u - v) = \left(-\frac{3}{5}\right) \cdot \frac{5}{13} + \frac{4}{5} \cdot \left(-\frac{12}{13}\right) = -\frac{15}{65} - \frac{48}{65} = -\frac{63}{65}.$$

b) Use the formula $\sin(u + v) = \sin u \cos v + \cos u \sin v$.

$$\sin(u + v) = \frac{4}{5} \cdot \frac{5}{13} + \left(-\frac{3}{5}\right) \cdot \left(-\frac{12}{13}\right) = \frac{20}{65} + \frac{36}{65} = \frac{56}{65}.$$

c) Use the formula $\tan(u - v) = \dfrac{\tan u - \tan v}{1 + \tan u \tan v}$.

$$\tan(u - v) = \frac{-\dfrac{4}{3} - \left(-\dfrac{12}{5}\right)}{1 + \left(-\dfrac{4}{3}\right) \cdot \left(-\dfrac{12}{5}\right)} = \frac{\dfrac{16}{15}}{1 + \dfrac{48}{15}} = \frac{\dfrac{16}{15}}{\dfrac{63}{15}} = \frac{16}{63}.$$

Practice Now. Given $\sin x = \dfrac{3}{5}$ and $\cos y = -\dfrac{2\sqrt{5}}{5}$ with $0 \leq x \leq \dfrac{\pi}{2}$ and $\dfrac{\pi}{2} \leq y \leq \pi$, find each trigonometric function value.

a) $\cos(x - y)$ **b)** $\sin(x + y)$ **c)** $\tan(x - y)$

Exercises 3.2

(1–11) Use identities to find each exact value. Do not use a calculator.

1. $\cos 75°$ **2.** $\tan 15°$

3. $\cos(-15°)$ **4.** $\cos(-105°)$

5. $\tan 75°$ **6.** $\sin 15°$

7. $\sin 345°$ **8.** $\tan(-105°)$

9. $\sin \dfrac{5\pi}{12}$ **10.** $\tan \dfrac{5\pi}{12}$

11. $\cos\left(-\dfrac{7\pi}{12}\right)$

(12–19) Write each function value in terms of the cofunction of a complementary angle.

12. $\tan 87°$ **13.** $\sin 15°$

14. $\cos \dfrac{\pi}{12}$ **15.** $\sin \dfrac{2\pi}{5}$

16. $\sin \dfrac{3\pi}{8}$ **17.** $\cot \dfrac{3\pi}{10}$

18. $\sec 46°24'$ **19.** $\tan 74°3'$

(20–27) Using the sum and difference identities, condense each of the following and express as a trigonometric function of a single angle.

20. $\cos 40° \cos 50° - \sin 40° \sin 50°$

21. $\cos 130° \cos 72° + \sin 130° \sin 72°$

22. $\sin 97° \cos 43° - \cos 97° \sin 43°$

23. $\dfrac{\tan 140° - \tan 60°}{1 + \tan 140° \tan 60°}$

24. $\cos \dfrac{7\pi}{9} \cos \dfrac{2\pi}{9} - \sin \dfrac{7\pi}{9} \sin \dfrac{2\pi}{9}$

25. $\cos \dfrac{\pi}{6} \cos \dfrac{\pi}{7} - \sin \dfrac{\pi}{6} \sin \dfrac{\pi}{7}$

26. $\sin \dfrac{\pi}{5} \cos \dfrac{2\pi}{3} - \cos \dfrac{\pi}{5} \sin \dfrac{2\pi}{3}$

27. $\dfrac{\tan \dfrac{\pi}{3} + \tan \dfrac{\pi}{4}}{1 - \tan \dfrac{\pi}{3} \tan \dfrac{\pi}{4}}$

28. If $\sin \theta = -\dfrac{3}{5}$ and θ is in the third quadrant, find $\cos\left(\theta + \dfrac{\pi}{3}\right)$.

(29–31) Verify the following identities.

29. $\sin(\pi - x) = \sin x$

30. $\sin\left(\dfrac{3\pi}{2} + x\right) = -\cos x$

31. $\sin\left(x + \dfrac{\pi}{2}\right) = \cos x$

(32–40) Use identities to write each expression in terms of $\sin x$ and $\cos x$.

32. $\cos(90° - x)$ **33.** $\cos(180° - x)$

34. $\sin(45° + x)$ **35.** $\cos\left(\dfrac{\pi}{2} + x\right)$

36. $\sin\left(\dfrac{5\pi}{6} - x\right)$ **37.** $\tan(60° + x)$

38. $\tan(x - 30°)$ **39.** $\tan\left(x + \dfrac{\pi}{6}\right)$

40. $\tan\left(\dfrac{\pi}{4} + x\right)$

(41–46) Use the given information to find $\cos(x + y)$, $\sin(x - y)$, and $\tan(x + y)$.

41. $\cos x = \dfrac{3}{5}$, $\sin y = \dfrac{5}{13}$, $0 \leq x, y \leq \dfrac{\pi}{2}$

42. $\cos x = -\dfrac{4}{5}$, $\sin y = \dfrac{3}{5}$, $\dfrac{\pi}{2} \leq x, y \leq \pi$

43. $\sin x = \dfrac{2}{3}$, $\sin y = -\dfrac{1}{3}$, $\dfrac{\pi}{2} \leq x \leq \pi, \dfrac{3\pi}{2} \leq y \leq 2\pi$

44. $\sin x = \dfrac{3}{5}$, $\sin y = -\dfrac{12}{13}$, $0 \leq x \leq \dfrac{\pi}{2}, \pi \leq y \leq \dfrac{3\pi}{2}$

45. $\cos x = -\dfrac{8}{17}$, $\cos y = -\dfrac{3}{5}$, $\pi \leq x, y \leq \dfrac{3\pi}{2}$

46. $\csc x = \dfrac{13}{5}$, $\tan y = -\dfrac{3}{4}$, $\dfrac{\pi}{2} \leq x \leq \pi, \dfrac{3\pi}{2} \leq y \leq 2\pi$

3.3   Double-Angle and Half-Angle Identities

In this section, we will learn the double-angle and half-angle formulas and practice how to solve problems using these formulas. We list a few types of problems we can solve.

1. Given a trigonometric function value of an angle and its quadrant information, we can use double-angle or half-angle formulas to find all trigonometric function values of double or half of the given angle.

2. Find trigonometric function values of angles that are half of speical angles $30° = \dfrac{\pi}{6}$ and $45° = \dfrac{\pi}{4}$.

3. Verify identities that contain multiple (half, double, or even triple) angles.

4. Reduce powers of trigonometric functions.

✎ **Double-Angle Identities**

$$\sin 2x = 2 \sin x \cos x \qquad\qquad \cos 2x = \cos^2 x - \sin^2 x$$

$$\tan 2x = \frac{2 \tan x}{1 - \tan^2 x} \qquad\qquad \cos 2x = 1 - 2 \sin^2 x$$

$$\cos 2x = 2 \cos^2 x - 1$$

✎ **Power-Reducing Identities**

$$\sin^2 x = \frac{1 - \cos 2x}{2} \qquad \cos^2 x = \frac{1 + \cos 2x}{2} \qquad \tan^2 x = \frac{1 - \cos 2x}{1 + \cos 2x}$$

✎ **Half-Angle Identities**

$$\sin \frac{x}{2} = \pm\sqrt{\frac{1 - \cos x}{2}}$$

$$\cos \frac{x}{2} = \pm\sqrt{\frac{1 + \cos x}{2}}$$

$$\tan \frac{x}{2} = \pm\sqrt{\frac{1 - \cos x}{1 + \cos x}} = \frac{1 - \cos x}{\sin x} = \frac{\sin x}{1 + \cos x}$$

The sign $+$ or $-$ depends on the quadrant in which the half-angle $\dfrac{x}{2}$ lies.

Example 1. Finding the Trigonometric Function Values of Double Angles

If $\sin \theta = -\dfrac{4}{5}$ and θ is in quadrant III, find the exact value of each expression.

  a) $\sin 2\theta$ **b)** $\cos 2\theta$ **c)** $\tan 2\theta$

Solution. To use double-angle identities, we first find $\cos \theta$ and $\tan \theta$ using fundamental identities.

$$\sin^2 \theta + \cos^2 \theta = 1 \qquad\qquad\qquad \text{(Pythagorean identity)}$$

$$\left(-\frac{4}{5}\right)^2 + \cos^2\theta = 1 \qquad\qquad \text{(Substitute given value.)}$$

$$\cos^2\theta = 1 - \left(-\frac{4}{5}\right)^2 = 1 - \frac{16}{25} = \frac{9}{25} \qquad \text{(Subtract the constant from both sides.)}$$

$$\cos\theta = \pm\sqrt{\frac{9}{25}} = \pm\frac{3}{5} \qquad\qquad \text{(Take square roots.)}$$

$$\cos\theta = -\frac{3}{5} \qquad\qquad \text{(For θ in quadrant III, $\cos\theta < 0$.)}$$

Next, using the reciprocal identity,

$$\tan\theta = \frac{\sin\theta}{\cos\theta} = \frac{-4/5}{-3/5} = \frac{4}{3}.$$

a) Use the sine double-angle identity.

$$\sin 2\theta = 2\sin\theta\cos\theta \qquad\qquad \text{(Double-angle identity for sine)}$$

$$= 2\left(-\frac{4}{5}\right)\left(-\frac{3}{5}\right) \qquad\qquad \text{(Substitute values.)}$$

$$= \frac{24}{25}$$

b) You may choose one of three cosine double-angle identities. Because we already know both $\cos\theta$ and $\sin\theta$, use the first of three double identities for cosine.

$$\cos 2\theta = \cos^2\theta - \sin^2\theta \qquad\qquad \text{(Double-angle identity for cosine)}$$

$$= \left(-\frac{3}{5}\right)^2 - \left(-\frac{4}{5}\right)^2 = -\frac{7}{25} \qquad \text{(Substitute values and simplify.)}$$

c) Use the double-angle identity for tangent.

$$\tan 2\theta = \frac{2\tan\theta}{1-\tan^2\theta} \qquad\qquad \text{(Double-angle identity for tangent)}$$

$$= \frac{2\cdot\left(\dfrac{4}{3}\right)}{1-\left(\dfrac{4}{3}\right)^2} \qquad\qquad \text{(Substitute the value.)}$$

$$= \frac{\dfrac{8}{3}}{1-\dfrac{16}{9}} = \frac{\dfrac{8}{3}}{-\dfrac{7}{9}} \qquad\qquad \text{(Simplify.)}$$

$$= \frac{8}{3}\cdot\left(-\frac{9}{7}\right) \qquad\qquad \text{(Divide by multiplying the reciprocal.)}$$

$$= -\frac{24}{7} \qquad\qquad \text{(Simplify.)}$$

Alternatively, we can also use the quotient identity $\tan 2\theta = \dfrac{\sin 2\theta}{\cos 2\theta}$ because we already computed $\sin 2\theta$ and $\cos 2\theta$ in previous parts.

$$\tan 2\theta = \frac{\sin 2\theta}{\cos 2\theta} = \frac{\dfrac{24}{25}}{-\dfrac{7}{25}} = -\frac{24}{7}$$

Practice Now. If $\sin\theta = -\dfrac{3}{5}$ and θ is in quadrant IV, find the exact value of each expression.

a) $\sin 2\theta$ **b)** $\cos 2\theta$ **c)** $\tan 2\theta$

Example 2. Finding Exact Values of Trigonometric Expressions Using Double-Angle Identities

Use double-angle identities to find exact values.

a) $1 - 2\sin^2\left(\dfrac{\pi}{8}\right)$ **b)** $\dfrac{2\tan 15°}{1 - \tan^2 15°}$

Solution. Use double-angle identities to simplify the expression, then evaluate it.

a) Note that the given expression is the right-hand side of the double-angle identity $\cos 2x = 1 - 2\sin^2 x$. By reversing sides, $1 - 2\sin^2 x = \cos 2x$. Now substitute the angle $\dfrac{\pi}{8}$.

$$1 - 2\sin^2\left(\frac{\pi}{8}\right) = \cos\left(2 \cdot \frac{\pi}{8}\right) = \cos\frac{\pi}{4} = \frac{\sqrt{2}}{2}.$$

b) Similarly, note that the given expression is the right-hand side of the identity $\tan 2x = \dfrac{2\tan x}{1 - \tan^2 x}$. By reversing sides, $\dfrac{2\tan x}{1 - \tan^2 x} = \tan 2x$. Now substitute $x = 15°$.

$$\frac{2\tan 15°}{1 - \tan^2 15°} = \tan(2 \cdot 15°) = \tan 30° = \frac{\sqrt{3}}{3}.$$

Practice Now. Find exact values.

a) $2\cos^2\left(\dfrac{\pi}{8}\right) - 1$ **b)** $\dfrac{2\tan 22.5°}{1 - \tan^2 22.5°}$

Example 3. Reducing Powers of Trigonometric Functions

Write an equivalent expression for $\cos^3 x$ that does not contain powers.

Solution. Use power-reducing identities to reduce powers. Because the given expression is a power of cosine, use the power-reducing identity for cosine.

$$\cos^3 x = \cos x \cos^2 x$$
$$= \cos x \cdot \frac{1 + \cos 2x}{2} \qquad \text{(Power-reducing identity for } \cos^2 x)$$
$$= \frac{1}{2}\cos x + \frac{1}{2}\cos x \cos 2x$$

Practice Now. Write an equivalent expression for $\sin^4 x$ that does not contain powers.

Example 4. Finding Trigonometric Function Values of Half-Angles

Use half-angle formulas to find the exact value of each expression.

a) $\sin \dfrac{\pi}{8}$ **b)** $\cos 165°$

Solution.

a) To use the half-angle identity, take two preliminary steps.

$$\sin \frac{x}{2} = \pm \sqrt{\frac{1 - \cos x}{2}}$$

① Determine the sign   ② Find the angle x

① To determine the sign, use the quadrant in which the angle lies. Because $\dfrac{\pi}{8}$ is in quadrant I, where all trigonometric functions are positive, so is $\sin \dfrac{\pi}{8}$. Use the $+$ sign.

② To find the angle x, solve $\dfrac{x}{2} = \dfrac{\pi}{8}$. Then get $x = 2 \times \dfrac{\pi}{8} = \dfrac{\pi}{4}$.

$$\sin \frac{\pi}{8} = +\sqrt{\frac{1 - \cos \frac{\pi}{4}}{2}} \qquad \text{(Use ① and ②.)}$$

$$= \sqrt{\frac{1 - \frac{\sqrt{2}}{2}}{2}}$$

$$= \sqrt{\frac{1 - \frac{\sqrt{2}}{2}}{2} \cdot \frac{2}{2}} \qquad \text{(Multiply } \frac{2}{2} \text{ to simplify double fraction.)}$$

$$= \sqrt{\frac{2 - \sqrt{2}}{4}} \qquad \text{(Simplify.)}$$

$$= \frac{\sqrt{2 - \sqrt{2}}}{\sqrt{4}} \qquad \text{(Split square root.)}$$

$$= \frac{\sqrt{2 - \sqrt{2}}}{2} \qquad \text{(Simplify the denominator.)}$$

b) Use the half-angle identity $\cos \dfrac{x}{2} = \pm \sqrt{\dfrac{1 + \cos x}{2}}$.

① Because $165°$ is in quadrant II, $\cos 165°$ is negative.

② Solve $\dfrac{x}{2} = 165°$ to get $x = 2 \times 165° = 330°$.

$$\cos 165° = -\sqrt{\frac{1 + \cos 330°}{2}}$$

$$= -\sqrt{\frac{1 + \frac{\sqrt{2}}{2}}{2}}$$

$$= -\sqrt{\frac{2 + \sqrt{2}}{4}} \qquad \text{(Simplify double fraction.)}$$

$$= -\frac{\sqrt{2 + \sqrt{2}}}{2} \qquad \text{(Simplify the denominator.)}$$

Practice Now. Use half-angle identities to find the exact value of each expression.

a) $\cos \dfrac{\pi}{8}$ **b)** $\tan 22.5°$

Example 5. Verifying Identities Using Double-Angle Identities

Verify the identity $\tan \theta = \dfrac{\sin 2\theta}{1 + \cos 2\theta}$.

Solution. We begin with the right-hand side, which is more complicated.

$$\begin{aligned}
\text{RHS} &= \frac{\sin 2\theta}{1 + \cos 2\theta} \\
&= \frac{2 \sin \theta \cos \theta}{1 + (2\cos^2 \theta - 1)} \qquad \text{(Double-angle identities for sine and cosine.)} \\
&= \frac{2 \sin \theta \cos \theta}{2 \cos^2 \theta} \qquad \text{(Simplify the denominator.)} \\
&= \frac{\sin \theta}{\cos \theta} \qquad \text{(Cancel the common factor.)} \\
&= \tan \theta \qquad \text{(Quotient identity)}
\end{aligned}$$

The result is exactly the same as the left-hand side. The identity is verified.

Practice Now. Verify the identity $\cot \theta = \dfrac{\sin 2\theta}{1 - \cos 2\theta}$.

Exercises 3.3

(1–6) Use the given information to find the exact values of the six trigonometric functions of 2θ.

1. $\sin \theta = \dfrac{3}{5}$, θ lies in quadrant I.

2. $\sin \theta = -\dfrac{5}{13}$, θ lies in quadrant IV.

3. $\cos \theta = \dfrac{2}{3}$, θ lies in quadrant IV.

4. $\cos \theta = -\dfrac{12}{13}$, θ lies in quadrant III.

5. $\tan \theta = 3$, θ lies in quadrant III.

6. $\sec \theta = \dfrac{7}{4}$, θ lies in quadrant I.

(7–13) Use double-angle identities to find exact values.

7. $2 \sin 15° \cos 15°$

8. $2 \sin \dfrac{\pi}{8} \cos \dfrac{\pi}{8}$

9. $\cos^2 15° - \sin^2 15°$

10. $1 - 2\sin^2\left(\dfrac{\pi}{12}\right)$

11. $2\cos^2 75° - 1$

12. $\dfrac{2 \tan 165°}{1 - \tan^2 165°}$

13. $\dfrac{2 \tan \dfrac{7\pi}{8}}{1 - \tan^2\left(\dfrac{7\pi}{8}\right)}$

(14–21) Verify each identity.

14. $(\sin x + \cos x)^2 = 1 + \sin 2x$

15. $(\sin x - \cos x)^2 = 1 - \sin 2x$

16. $\cos^2 x = \cos 2x + \sin^2 x$

17. $\tan \dfrac{x}{2} = \dfrac{1 - \cos x}{\sin x}$

18. $\sin 3x = 3 \sin x - 4 \sin^3 x$

19. $\cos 3x = 4\cos^3 x - 3\cos x$

20. $\sin 4\theta = 4\sin\theta\cos^3\theta - 4\sin^3\theta\cos\theta$

21. $\cos 4\theta = 8\cos^4\theta - 8\cos^2\theta + 1$

(22–28) Use half-angle identities to find exact values.

22. $\sin 15°$

24. $\sin 105°$

26. $\tan\dfrac{7\pi}{8}$

28. $\sin 157.5°$

23. $\cos 15°$

25. $\cos 75°$

27. $\tan 75°$

3.4   Product-to-Sum and Sum-to-Product Identities

In this section, we will learn how to change the form of some trigonometric expressions from products to sums and sums back to products.

Product-to-Sum Identities

$$\sin x \cos y = \frac{1}{2}[\sin(x+y) + \sin(x-y)]$$

$$\cos x \sin y = \frac{1}{2}[\sin(x+y) - \sin(x-y)]$$

$$\cos x \cos y = \frac{1}{2}[\cos(x+y) + \cos(x-y)]$$

$$\sin x \sin y = \frac{1}{2}[\cos(x-y) - \cos(x+y)]$$

Sum-to-Product Identities

$$\sin x + \sin y = 2\sin\left(\frac{x+y}{2}\right)\cos\left(\frac{x-y}{2}\right)$$

$$\sin x - \sin y = 2\cos\left(\frac{x+y}{2}\right)\sin\left(\frac{x-y}{2}\right)$$

$$\cos x + \cos y = 2\cos\left(\frac{x+y}{2}\right)\cos\left(\frac{x-y}{2}\right)$$

$$\cos x - \cos y = -2\sin\left(\frac{x+y}{2}\right)\sin\left(\frac{x-y}{2}\right)$$

▶ Product-to-Sum Identities

We can use the product-to-sum identities to rewrite products of sines and/or cosines as sums or differences. The proofs of these identities are simple and very similar. For example, to prove the first identity

$$\sin x \cos y = \frac{1}{2}[\sin(x+y) + \sin(x-y)],$$

we add the sum and difference identities of sine:

$$
\begin{array}{rcl}
\sin(x+y) & = & \sin x \cos y + \cos x \sin y \\
\sin(x-y) & = & \sin x \cos y - \cos x \sin y \\
\hline
\sin(x+y) + \sin(x-y) & = & 2\sin x \cos y
\end{array}
$$

Switching sides and multiplying both sides by $\frac{1}{2}$, we have

$$\sin x \cos y = \frac{1}{2}[\sin(x + y) + \sin(x - y)]$$

Example 1. Using the Product-to-Sum Identities to Convert Expressions

Express each of the following products as a sum or difference.

a) $\sin 6x \sin 4x$ **b)** $\cos 5x \sin 3x$

Solution. Apply appropriate formulas.

a) $\sin 6x \sin 4x = \dfrac{1}{2}[\cos(6x - 4x) - \cos(6x + 4x)]$ (Product-to-Sum Identity)

$$= \frac{1}{2}(\cos 2x - \cos 10x)$$

b) $\cos 5x \sin 3x = \dfrac{1}{2}[\sin(5x + 3x) - \sin(5x - 3x)]$ (Product-to-Sum Identity)

$$= \frac{1}{2}(\sin 8x - \sin 2x)$$

Practice Now. Express each of the following products as a sum or difference.

a) $\sin 7x \cos 3x$ **b)** $\cos 7x \cos 3x$

Example 2. Using the Product-to-Sum Identities to Find Exact Values of Expressions

Find the exact value of $\cos 105° \cos 15°$.

Solution. Use the identity $\cos x \cos y = \dfrac{1}{2}[\cos(x + y) + \cos(x - y)]$.

$\cos 105° \cos 15° = \dfrac{1}{2}[\cos(105° + 15°) + \cos(105° - 15°)]$ (Set $x = 105°$ and $y = 15°$)

$$= \frac{1}{2}(\cos 120° + \cos 90°)$$

$= \dfrac{1}{2}(-\dfrac{1}{2} + 0)$ (Because $\cos 120° = -\dfrac{1}{2}$; $\cos 90° = 0$.)

$$= -\frac{1}{4}$$

Practice Now. Find the exact values.

a) $\sin 75° \cos 15°$ **b)** $\sin 165° \sin 15°$

▶ Sum-to-Product Identities

We can use sum-to-product identities to write the sum or difference of sines and/or cosines as products. The proofs of sum-to-product identities are simple and similar. For example, to prove the first identity

$$\sin x + \sin y = 2 \sin \frac{x + y}{2} \cos \frac{x - y}{2},$$

we use the product-to-sum identity $\sin x \cos y = \dfrac{1}{2}[\sin(x+y) + \sin(x-y)]$.

$$2\sin\frac{x+y}{2}\cos\frac{x-y}{2} = 2\cdot\frac{1}{2}\left[\sin\left(\frac{x+y}{2}+\frac{x-y}{2}\right) + \sin\left(\frac{x+y}{2}+\frac{x-y}{2}\right)\right]$$
$$= \sin x + \sin y \qquad\qquad\qquad\qquad\qquad \text{(Simplify)}$$

Example 3. Using the Sum-to-Product Identities to Convert Expressions

Express the sum as a product.

a) $\sin 7x + \sin 3x$ **b)** $\cos 5x - \cos x$

Solution.   Use the appropriate sum-to-product identity for each one.

a)
$$\sin 7x + \sin 3x = 2\sin\frac{7x+3x}{2}\cos\frac{7x-3x}{2} \quad \text{(Sum-to-Product Identity)}$$
$$= 2\sin 5x \cos 2x$$

b)
$$\cos 5x - \cos x = -2\sin\frac{5x+x}{2}\sin\frac{5x-x}{2} \quad \text{(Sum-to-Product Identity)}$$
$$= -2\sin 3x \sin 2x$$

Practice Now.   Express the sum as a product.

a) $\cos 6x + \cos 2x$ **b)** $\sin 4x - \sin 2x$

Example 4. Using the Sum-to-Product Identities to Verify Identities

Verify the identity $\dfrac{\cos 6x - \cos 4x}{\sin 6x + \sin 4x} = -\tan x$.

Solution.   We use a sum-to-product identity for the numerator and another one for the denominator:

$$\frac{\cos 6x - \cos 4x}{\sin 6x + \sin 4x} = \frac{-2\sin\dfrac{6x+4x}{2}\sin\dfrac{6x-4x}{2}}{2\sin\dfrac{6x+4x}{2}\cos\dfrac{6x-4x}{2}}$$
$$= -\frac{\sin 5x \sin x}{\sin 5x \cos x} \qquad\qquad \text{(Simplify)}$$
$$= -\frac{\sin x}{\cos x}$$
$$= -\tan x. \qquad\qquad\qquad \text{(Quotient identity for tangent)}$$

Practice Now.   Verify the identity $\dfrac{\sin 5x + \sin 7x}{\cos 5x - \cos 7x} = \cot x$.

Exercises 3.4

(1–8) Use the product-to-sum identities to rewrite each expression as the sum or difference and simplify.

1. $\sin 2x \sin 3x$

2. $\sin 4x \cos 6x$

3. $\cos x \sin x$

4. $\cos 5x \cos 7x$

5. $\sin 35° \cos 5°$

6. $\cos 108° \cos 52°$

7. $\sin \dfrac{7\pi}{12} \sin \dfrac{\pi}{12}$

8. $\cos \dfrac{2\pi}{5} \sin \dfrac{\pi}{5}$

(9–13) Find the exact value.

9. $\sin 105° \cos 15°$

10. $\sin 7.5° \sin 52.5°$

11. $\cos 67.5° \cos 22.5°$

12. $\cos 52.5° \sin 7.5°$

13. $\cos \left(-\dfrac{5\pi}{24} \right) \sin \left(\dfrac{\pi}{24} \right)$

(14–22) Use the sum-to-product identities to rewrite each expression as a product and simplify.

14. $\sin 50° - \sin 40°$

15. $\sin 40° + \sin 20°$

16. $\cos 50° + \cos 10°$

17. $\cos 25° - \cos 65°$

18. $\sin \dfrac{3\pi}{7} + \cos \dfrac{4\pi}{7}$

19. $\sin 3x + \sin 5x$

20. $\sin 7x - \sin x$

21. $\cos 3x + \cos 5x$

22. $\cos 4x - \cos 6x$

(23–26) Verify each identity

23. $\dfrac{\sin 2x + \sin 6x}{\cos 2x + \cos 6x} = \tan 4x$

24. $\dfrac{\cos x + \cos 3x}{\sin x + \sin 3x} = \cot 2x$

25. $\dfrac{\cos 7x - \cos 5x}{\cos 7x + \cos 5x} = -\tan 6x \tan x$

26. $\dfrac{\sin 3x + \sin x}{\sin 3x - \sin x} = \tan 2x \cot x$

3.5   Trigonometric Equations

We will first learn how to solve basic trigonometric equations involving one single trigonometric function in the form of
$$\sin x = a, \quad \cos x = a, \quad \text{or} \quad \tan x = a, \quad a : \text{constant}.$$
Then we will proceed to study more complicated equations that require techniques used for linear and quadratic equations.

▶ Basic Trigonometric Equations

Sine and Cosine Equations

Solving a trigonometric equation is finding "all" angles that give the specified trigonometric value. One way to find an angle is to use an inverse trigonometric function. But that approach is not enough because an inverse function gives at most one solution, and we want all solutions.

For example, there are infinitely many solutions of the sine equation $\sin x = \dfrac{1}{2}$, as follows.

$$\sin x = \frac{1}{2} \quad \Longrightarrow \quad x = \ldots, -\frac{11\pi}{6}, -\frac{7\pi}{6}, \frac{\pi}{6}, \frac{5\pi}{6}, \frac{13\pi}{6}, \frac{17\pi}{6}, \ldots$$

You can see them in the graph of the sine function. There are **principal solutions** in the first period $[0, 2\pi)$, and there are other solutions that repeat every period.

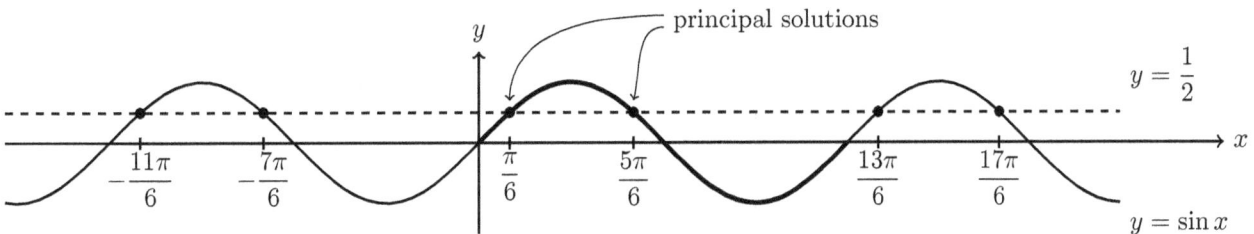

When there are infinitely many solutions, we write them as follows (adding the expression $2k\pi$ to the principal solutions) and call them **general solutions**,

$$\frac{\pi}{6} + 2k\pi, \quad \frac{5\pi}{6} + 2k\pi, \qquad k: \text{integer},$$

because these expressions generate all the solutions when we substitute k with integers.

k	$\cdots$	-1	0	1	2	$\cdots$
$\frac{\pi}{6} + 2k\pi$	$\cdots$	$-\frac{11\pi}{6}$	$\frac{\pi}{6}$	$\frac{13\pi}{6}$	$\frac{25\pi}{6}$	$\cdots$
$\frac{5\pi}{6} + 2k\pi$	$\cdots$	$-\frac{7\pi}{6}$	$\frac{5\pi}{6}$	$\frac{17\pi}{6}$	$\frac{29\pi}{6}$	$\cdots$

Solving a basic sine or cosine equation for general solutions is done in two steps.

Step 1. Find principal solutions in the interval $[0, 2\pi)$ or $[0°, 360°)$.

Step 2. Add $2k\pi$ or $360°k$ for general solutions.

Here is how we got principal solutions. We used the reference angle and sketch to find them, as follows.

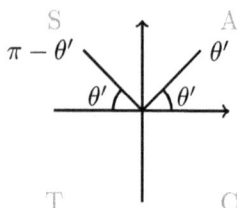

Because we are solving an equation with positive sine value $\frac{1}{2}$, solutions are in quadrants I and II, where sine is positive. The reference angle is obtained by taking the inverse sine of the absolute value of the given sine value.

$$\theta' = \sin^{-1}\left|\frac{1}{2}\right| = 30° = \frac{\pi}{6}$$

Then the principal solutions are $\frac{\pi}{6}$ and $\pi - \frac{\pi}{6} = \frac{5\pi}{6}$ in quadrants I and II.

Example 1. Solving Basic Sine and Cosine Equations

Find all solutions of the equation.

a) $\sin\theta = \dfrac{\sqrt{3}}{2}$ **b)** $\cos x = -\dfrac{1}{2}$ **c)** $\sin t = 3$ **d)** $\cos x = 0$

Solution.

a) Because the given sine value is positive, the solutions are in quadrants I and II.
The reference angle is

$$\theta' = \sin^{-1}\left|\frac{\sqrt{3}}{2}\right| = \sin^{-1}\frac{\sqrt{3}}{2} = 60° = \frac{\pi}{3}.$$

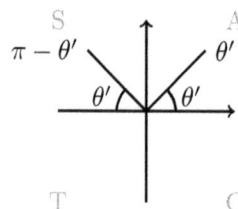

Thus, the principal solutions are $\dfrac{\pi}{3}$ and $\pi - \dfrac{\pi}{3} = \dfrac{2\pi}{3}$.

Add $2\pi k$ for the general solutions.

$$\theta = \frac{\pi}{3} + 2k\pi, \quad \frac{2\pi}{3} + 2k\pi, \qquad k : \text{integer}$$

b) Because the given cosine value is negative, the solutions are in quadrants II and III.

The reference angle is

$$\theta' = \cos^{-1}\left|-\frac{1}{2}\right| = \cos^{-1}\frac{1}{2} = 60° = \frac{\pi}{3}.$$

Thus, the principal solutions are $\pi - \frac{\pi}{3} = \frac{2\pi}{3}$ and $\pi + \frac{\pi}{3} = \frac{4\pi}{3}$.

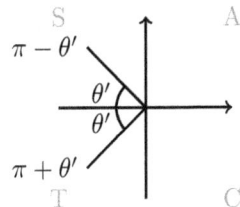

Add $2\pi k$ for the general solutions.

$$x = \frac{2\pi}{3} + 2k\pi, \quad \frac{4\pi}{3} + 2k\pi, \qquad k : \text{integer}$$

c) This equation has no solution because the range of the sine function is $[-1, 1]$ and 3 is out of the range. In fact, when you try to compute $\sin^{-1} 3$ with the calculator for the reference angle, it shows an error.

d) This equation is a little different because 0 is neither positive nor negative. So, we can't determine quadrants. Instead, we use the graph of the cosine function to locate zeros.

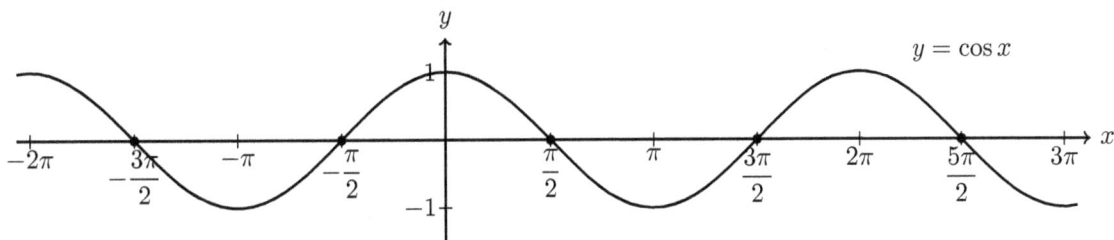

Notice that the solutions repeat every π. So, the answer is

$$\frac{\pi}{2} + k\pi, \qquad k : \text{integer}.$$

Practice Now.   Solve the equation.

a) $\sin\theta = -\frac{1}{2}$ 　　　　**b)** $\cos x = \frac{\sqrt{2}}{2}$ 　　　　**c)** $\cos u = -\frac{5}{4}$ 　　　　**d)** $\sin t = 0$

Tangent Equations

Because the period of the tangent function is π, the solutions of tangent equations repeat every π. The principal solution of the tangent equation lies in the interval $\left(-\frac{\pi}{2}, \frac{\pi}{2}\right)$, and it is obtained by taking the inverse tangent of the given tangent value. Therefore, the solutions of the equation $\tan x = a$ are $x = \theta + k\pi$ where $\theta = \tan^{-1} a$. You can see the solutions in the graph of the tangent function.

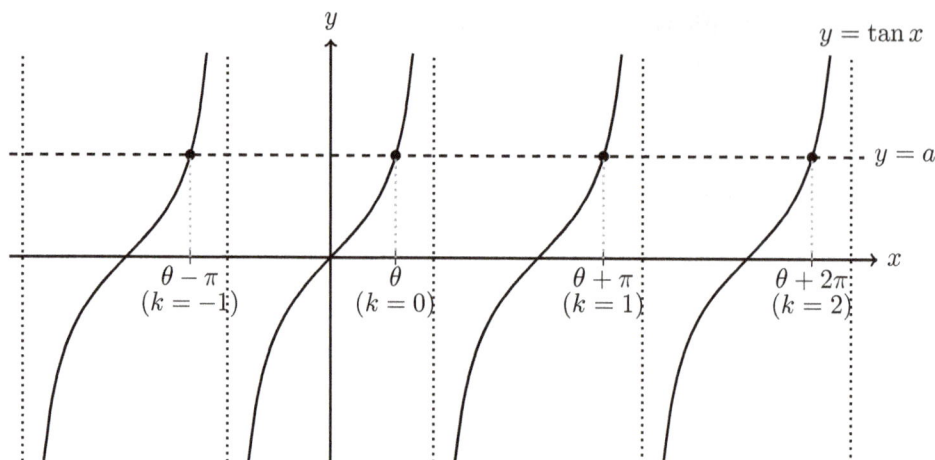

Example 2. Solving Basic Tangent Equations

Find the solutions of the equation.

a) $\tan\theta = 1$

b) $\tan\theta = -\sqrt{3}$

Solution.

a) Because $\tan^{-1}1 = 45° = \dfrac{\pi}{4}$, the solutions are $\dfrac{\pi}{4} + k\pi$ where k is an integer.

b) Because $\tan^{-1}(-\sqrt{3}) = -60° = -\dfrac{\pi}{3}$, the solutions are $-\dfrac{\pi}{3} + k\pi$ where k is an integer.

Practice Now.   Solve the equation.

a) $\tan x = -1$

b) $\tan u = \dfrac{\sqrt{3}}{3}$

▶ More Trigonometric Equations

In this subsection, we discuss more examples that require additional steps.

Example 3. Solving Trigonometric Equations Involving Multiple Angles

a) Solve the equation $\sin 3x = 0$. Express solutions in degrees.

b) Find the solutions that are in the interval $[0°, 360°)$.

Solution.

a) First, substitute $\theta = 3x$. Then the equation becomes $\sin\theta = 0$, a basic sine equation. Solve the equation for θ. The zeros of the sine function are at every multiple of π ($\ldots$, -2π, $-\pi$, 0, π, 2π, 3π, $\ldots$), as seen in the sine graph below.

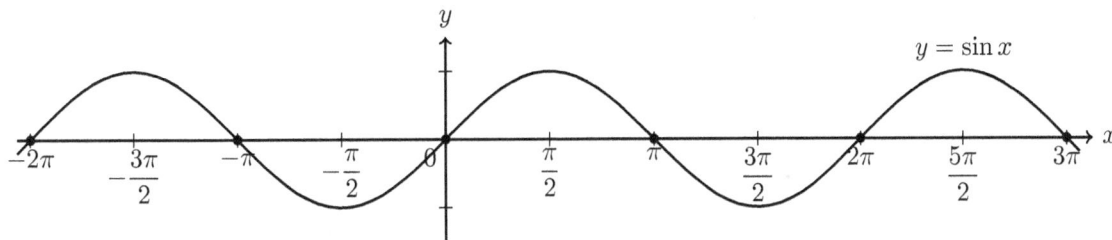

Therefore, $\theta = k\pi$ or $\theta = 180°k$ in degrees for integers k. Now substitute $\theta = 3x$ back and solve for x.

$$\theta = 180°k$$
$$3x = 180°k \qquad \text{(Substitute back.)}$$
$$x = 60°k \qquad \text{(Divide by 3.)}$$

b) Substitute integers for k in the formula for x obtained in the previous part and make a table of values.

k	$\cdots$	-1	0	1	2	3	4	5	6	$\cdots$
$60°k$	$\cdots$	$-60°$	$0°$	$60°$	$120°$	$180°$	$240°$	$300°$	$360°$	$\cdots$

Then find the solutions in the interval $[0°, 360°)$. The solutions are as follows.

$$0°, 60°, 120°, 180°, 240°, \text{ and } 300°.$$

Practice Now.

   a) Solve the equation $\cos 2x = 1$ and express the solutions in radians.

   b) Find the solutions that are in the interval $[0, 2\pi)$.

Example 4. Solving Linear Trigonometric Equations

Solve the equation $2\cos\left(x + \dfrac{\pi}{4}\right) - 1 = 0$. Then find all solutions in the interval $[0, 2\pi)$.

Solution.   First, substitute the angle, $\theta = x + \dfrac{\pi}{4}$. Then solve the equation for $\cos\theta$.

$$2\cos\theta - 1 = 0 \qquad \text{(Substitution)}$$
$$2\cos\theta = 1 \qquad \text{(Add 1 to both sides.)}$$
$$\cos\theta = \frac{1}{2} \qquad \text{(Divide both sides by 2.)}$$

Next, solve the equation for θ. Because the cosine value is positive, the solutions are in quadrants I and IV.

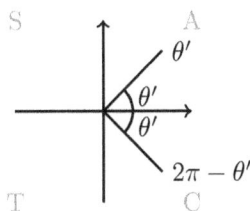

The reference angle is

$$\theta' = \cos^{-1}\left|\frac{1}{2}\right| = \cos^{-1}\frac{1}{2} = 60° = \frac{\pi}{3}.$$

Thus, the principal solutions are $\dfrac{\pi}{3}$ and $2\pi - \dfrac{\pi}{3} = \dfrac{5\pi}{3}$.

Then the general solutions are $\theta = \dfrac{\pi}{3} + 2k\pi$ and $\dfrac{5\pi}{3} + 2k\pi$.

Finally, substitute the angle back, and solve for x.

$$\theta = \frac{\pi}{3} + 2k\pi$$

$$x + \frac{\pi}{4} = \frac{\pi}{3} + 2k\pi \qquad \text{(Substitute back.)}$$

$$x = \frac{\pi}{3} - \frac{\pi}{4} + 2k\pi \qquad \text{(Subtract } \pi/4.)$$

$$x = \frac{\pi}{12} + 2k\pi \qquad \text{(Simplify fractions.)}$$

$$\theta = \frac{5\pi}{3} + 2k\pi$$

$$x + \frac{\pi}{4} = \frac{5\pi}{3} + 2k\pi \qquad \text{(Substitute back.)}$$

$$x = \frac{5\pi}{3} - \frac{\pi}{4} + 2k\pi \qquad \text{(Subtract } \pi/4)$$

$$x = \frac{17\pi}{12} + 2k\pi \qquad \text{(Simplify fractions.)}$$

The solutions in $[0, 2\pi)$ are $\frac{\pi}{12}$ and $\frac{17\pi}{12}$.

Practice Now. Solve the equation. Then find all solutions in the interval $[0, 2\pi)$.

a) $4\sin\left(3x - \frac{2\pi}{3}\right) + 2 = 0$

b) $\sqrt{3}\tan\left(2x - \frac{\pi}{2}\right) + 1 = 0$

Example 5. Solving Trigonometric Equations by Factoring

Solve the equation $6\sin^2\theta - \sin\theta - 1 = 0$, and write solutions in degrees.

Solution. The equation is quadratic in $\sin\theta$. Substitute $X = \sin\theta$. Then the equation is $6X^2 - X - 1 = 0$. Solve the quadratic equation by factoring.

$$6X^2 - X - 1 = 0$$

$$(2X - 1)(3X + 1) = 0$$

$$X = \frac{1}{2} \quad \text{or} \quad X = -\frac{1}{3}$$

Now substitute back $\sin\theta$ and get two sine equations.

$$\sin\theta = \frac{1}{2} \quad \text{or} \quad \sin\theta = -\frac{1}{3}$$

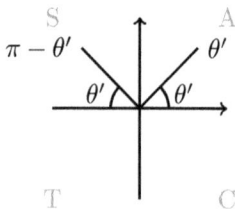

The first equation has solutions in quadrants I and II with reference angle

$$\sin^{-1}\frac{1}{2} = 30°.$$

Therefore, principal solutions are $30°$ and $180° - 30° = 150°$. Thus,

$$x = 30° + 360°k, \quad 150° + 360°k, \qquad k: \text{integer}.$$

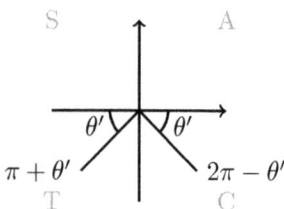

The second equation has solutions in quadrants III and IV with reference angle

$$\sin^{-1}\left(-\frac{1}{3}\right) \approx 19.5°.$$

Therefore, principal solutions are $180° + 19.5° = 199.5°$ and $360° - 19.5° = 340.5°$. Thus,

$$x \approx 199.5° + 360°k, \quad 340.5° + 360°k, \qquad k: \text{integer}.$$

Practice Now. Solve the equation, and write solutions in degrees.

a) $\cos x \tan x - 2 \tan x = 0$

b) $2 \cos^2 x + \cos x - 1 = 0$

Exercises 3.5

(1–7) Find all solutions of each equation. Write solutions in radians.

1. $\sin x = \dfrac{\sqrt{2}}{2}$

2. $\cos x = -\dfrac{\sqrt{3}}{2}$

3. $\tan x = \dfrac{\sqrt{3}}{3}$

4. $3 \cos x - 1 = \cos x$

5. $2 \sin \theta + 1 = 4 \sin \theta$

6. $\sec \theta = \dfrac{3}{\sqrt{2}}$

7. $\cot \theta = 0$

(8–12) Find the solutions of each equation in the interval $[0, 2\pi)$.

8. $\sin \left(x - \dfrac{\pi}{4} \right) = -\dfrac{1}{2}$

9. $\cos \left(x + \dfrac{\pi}{6} \right) = \dfrac{\sqrt{2}}{2}$

10. $\sin 2x = -\dfrac{\sqrt{2}}{2}$

11. $\sqrt{3} \tan \dfrac{3x}{4} = 1$

12. $\sin \left(3x - \dfrac{\pi}{3} \right) = -1$

(13–17) Find all solutions of each equation in the interval $[0, 2\pi)$.

13. $(\tan x - 1)(2 \sin x + 1) = 0$

14. $(2 \sin x + \sqrt{3})(2 \cos x + 1) = 0$

15. $\sin^2 x - 2 \sin x - 3 = 0$

16. $4 \cos^2 \theta - 1 = 0$

17. $\csc^2 x - 2 = 0$

(18–21) Use trigonometric identities to solve each equation, and find all solutions in the interval $[0°, 360°)$.

18. $2 \sin x \cos x = \dfrac{1}{2}$

19. $\cos^2 x - \sin^2 x = \dfrac{\sqrt{3}}{2}$

20. $\sin 2x \cos x + \cos 2x \sin x = 1$

21. $\cos x \cos 2x - \sin x \sin 2x = 1$

Chapter 4

APPLICATIONS OF TRIGONOMETRIC FUNCTIONS

4.1 The Law of Sines

▶ The Law of Sines

In this section, the angles and sides of a triangle $\triangle ABC$ are labeled as follows. The angles are labeled A, B and C, and their opposite sides are labeled a, b, and c, respectively.

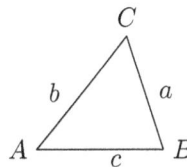

The law of sines describe the relationships between the three sides and three angles of a triangle as follows.

> ✎ **The Law of Sines**
>
> $$\frac{a}{\sin A} = \frac{b}{\sin B} = \frac{c}{\sin C} \qquad \frac{\sin A}{a} = \frac{\sin B}{b} = \frac{\sin C}{c}$$

The process of finding unknown sides and angles of a triangle is called **solving a triangle**. We will learn how to use the law of sines to solve three types of triangles.
- **ASA triangles**: Two angles and the included side are known.
- **AAS triangles**: Two angles and a nonincluded side are known.
- **SSA triangles**: Two sides and a nonincluded angle are known.

▶ Solving Oblique Triangles

Example 1. Solving an ASA Triangle

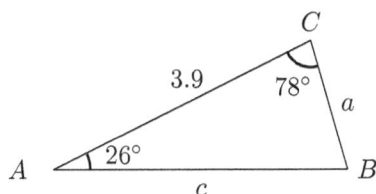

Solve the triangle shown on the left with $A = 26°$, $C = 78°$, and $b = 3.9$ inches. Round lengths of sides to the nearest tenth.

Solution.   First, find the angle B using the fact that the sum of all three angles of a triangle is $180°$. If two angles are given, the third angle is $180°$ minus two angles. Thus,

$$B = 180° - A - C = 180° - 26° - 78° = 76°.$$

Then use the law of sines to find sides a and c. On the third line below, multiply both sides by the denominator on the left to simplify equations.

$$\frac{a}{\sin A} = \frac{b}{\sin B}$$
$$\frac{a}{\sin 26°} = \frac{3.9}{\sin 76°}$$
$$\cancel{\sin 26°} \cdot \frac{a}{\cancel{\sin 26°}} = \frac{3.9}{\sin 76°} \cdot \sin 26°$$
$$a = \frac{3.9 \sin 26°}{\sin 76°}$$
$$\approx 1.8 \text{ inches}$$

$$\frac{c}{\sin C} = \frac{b}{\sin B}$$
$$\frac{c}{\sin 78°} = \frac{3.9}{\sin 76°}$$
$$\cancel{\sin 78°} \cdot \frac{c}{\cancel{\sin 78°}} = \frac{3.9}{\sin 76°} \cdot \sin 78°$$
$$c = \frac{3.9 \sin 78°}{\sin 76°}$$
$$\approx 3.9 \text{ inches}$$

Practice Now.   Solve the triangle $\triangle ABC$ with $A = 48°$, $B = 67°$, and $c = 5.1$ inches. Round lengths of sides to the nearest tenth.

Example 2. Solving an AAS Triangle

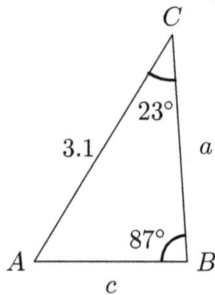

Solve the triangle $\triangle ABC$ with $B = 87°$, $C = 23°$, and $b = 3.1$ centimeters. Round lengths of sides to the nearest tenth.

Solution.   Use exactly the same method as in the ASA case. First, find the unknown angle A by subtracting two given angles from $180°$.

$$A = 180° - B - C = 180° - 87° - 23° = 70°$$

Then use the law of sines to solve for a and c.

$$\frac{a}{\sin A} = \frac{b}{\sin B}$$
$$\frac{a}{\sin 70°} = \frac{3.1}{\sin 87°}$$
$$\cancel{\sin 70°} \cdot \frac{a}{\cancel{\sin 70°}} = \frac{3.1}{\sin 87°} \cdot \sin 70°$$
$$a = \frac{3.1 \sin 70°}{\sin 87°}$$
$$\approx 2.9 \text{ cm}$$

$$\frac{c}{\sin C} = \frac{b}{\sin B}$$
$$\frac{c}{\sin 23°} = \frac{3.1}{\sin 87°}$$
$$\cancel{\sin 23°} \cdot \frac{c}{\cancel{\sin 23°}} = \frac{3.1}{\sin 87°} \cdot \sin 23°$$
$$c = \frac{3.1 \sin 23°}{\sin 87°}$$
$$\approx 1.2 \text{ cm}$$

Practice Now.   Solve the triangle $\triangle ABC$ with $B = 35°$, $C = 42°$, and $c = 10$ centimeters.

▶ The Ambiguous Case

If two sides and an angle opposite one of them are given (SSA), there might be no, or one, or even two triangles satisfying the given information. We call this the **ambiguous case**. For example, suppose we are given a, b, and A, and let $h = b \sin A$, which is the height in the figures below. There are five cases, depending on how long a is relative to b and h.

Case 1. If $a < h$, there is no triangle.

Case 2. If $a = h$, there is one right triangle.

Case 3. If $h < a < b$, there are two triangles.

Case 4. If $a = b$, there is one isosceles triangle.

Case 5. If $a > b$, there is one triangle.

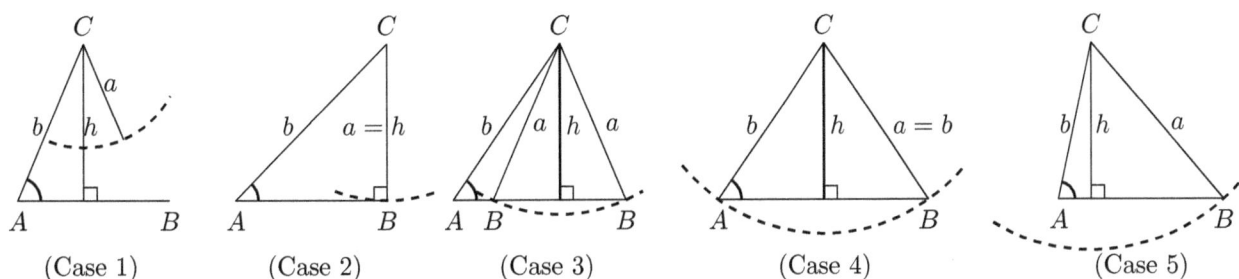

(Case 1)   (Case 2)   (Case 3)   (Case 4)   (Case 5)

Example 3. Solving an SSA Triangle: One Solution

Solve the triangle $\triangle ABC$ with $A = 46°$, $a = 22$, and $b = 14$. Round your answers to the nearest tenth.

Solution.   Because $a > b$, there is only one solution.

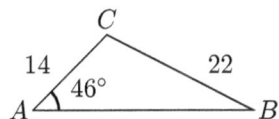

$$\frac{\sin B}{b} = \frac{\sin A}{a} \qquad \text{(The law of sines)}$$

$$\frac{\sin B}{14} = \frac{\sin 46°}{22} \qquad \text{(Substitute values.)}$$

$$\sin B = \frac{14 \sin 46°}{22} \qquad \text{(Multiply 14 on both sides.)}$$

$$\approx 0.4578 \qquad \text{(Approximate.)}$$

There are two angles B between $0°$ and $180°$ for which $\sin B \approx 0.4578$:

$$B \approx \sin^{-1} 0.4578 \approx 27.2° \quad \text{or} \quad B \approx 180° - 27.2° = 152.8°.$$

But B cannot be $152.8°$ because it would mean $C \approx 180° - 46° - 152° = -18.8°$ and C cannot be negative. Thus, $B \approx 27.2°$ and $C \approx 180° - 46° - 27.2° = 106.8°$. Then use the law of sine again to solve for c.

$$\frac{c}{\sin C} = \frac{b}{\sin B}$$

$$c = \frac{b \sin C}{\sin B} \qquad \text{(Solve for } c\text{.)}$$

$$\approx \frac{14 \sin 106.8°}{\sin 27.2°} \qquad \text{(Substitute values.)}$$

$$\approx 29.3$$

Practice Now.　Solve the triangle $\triangle ABC$ with $A = 57°$, $a = 12$, and $b = 9$. Round your answers to the nearest tenth.

Example 4. Solving an SSA Triangle: No Solution

Solve the triangle $\triangle ABC$ with $A = 38°$, $a = 4$, and $b = 7$. Round your answers to the nearest tenth.

Solution.　There is no solution. One way to see it is to compare lengths.

$$4 = a < h = b \sin A = 7 \sin 38° \approx 4.3$$

Another way is to try using the law of sines to find B.

$$\frac{\sin B}{b} = \frac{\sin A}{a}$$

$$\frac{\sin B}{7} = \frac{\sin 38°}{4}$$

$$\sin B = \frac{7 \sin 38°}{4}$$

$$\approx 1.077 > 1$$

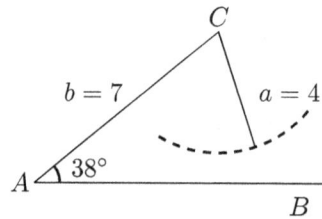

The sine equation has no solution because the sine value can't be greater than 1. Thus, no triangle with the given parts exists.

Example 5. Solving an SSA Triangle: One Right Triangle

Solve the triangle $\triangle ABC$ with $A = 30°$, $a = 6$, and $b = 12$. Round your answers to the nearest tenth.

Solution.　Because $h = a \sin A = 12 \sin 30° = 6$, so $a = h$, the triangle must be a right triangle.

$$\frac{\sin B}{b} = \frac{\sin A}{a}$$

$$\frac{\sin B}{12} = \frac{\sin 30°}{6}$$

$$\sin B = \frac{12 \sin 30°}{6}$$

$$= 1$$

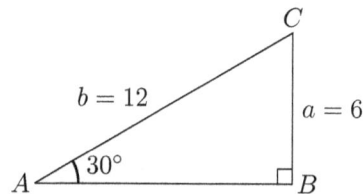

There is one angle $B = 90°$ between $0°$ and $180°$ with sine value 1. Then find the angle C.

$$C = 180° - A - B = 180° - 30° - 90° = 60°$$

Then we use the law of sines again to find c (we can also use the Pythagorean theorem to find c because the triangle is a right triangle).

$$\frac{c}{\sin C} = \frac{a}{\sin A}$$

$$\frac{c}{\sin 60°} = \frac{6}{\sin 30°}$$

$$c = \frac{6 \sin 60°}{\sin 30°}$$

$$\approx 10.4$$

Practice Now. Solve the triangle $\triangle ABC$ with $b = 6$, $c = 6\sqrt{2}$, and $B = 45°$.

Example 6. Solving an SSA Triangle: Two Solutions

Solve the triangle $\triangle ABC$ with $A = 52°$, $a = 25$ inches, and $b = 28$ inches. Round your answers to the nearest tenth.

Solution. Because $h = b \sin A = 28 \sin 52° \approx 22.1$, and $h \approx 22.1 < a = 25 < b = 28$, there are two triangles with the given parts. Use the law of sines to find B first.

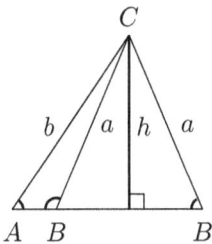

$$\frac{\sin B}{b} = \frac{\sin A}{a}$$

$$\frac{\sin B}{28} = \frac{\sin 52°}{25}$$

$$\sin B = \frac{28 \sin 52°}{25} \approx 0.8826$$

There are two angles between $0°$ and $180°$ with sine value 0.8826. Find B and solve for c using the law of sines again.

$$B \approx \sin^{-1} 0.8826$$
$$\approx 62.0°$$
$$C = 180° - A - B$$
$$\approx 180° - 52° - 62°$$
$$= 66°$$

$$\frac{c}{\sin C} = \frac{a}{\sin A}$$
$$\frac{c}{\sin 66°} = \frac{25}{\sin 52°}$$
$$c = \frac{25 \sin 66°}{\sin 52°}$$
$$\approx 29.0$$

$$B \approx 180° - \sin^{-1} 0.8826$$
$$\approx 180° - 62° = 118°$$
$$C = 180° - A - B$$
$$\approx 180° - 52° - 118°$$
$$= 10°$$

$$\frac{c}{\sin C} = \frac{a}{\sin A}$$
$$\frac{c}{\sin 10°} = \frac{25}{\sin 52°}$$
$$c = \frac{25 \sin 10°}{\sin 52°}$$
$$\approx 5.5$$

Practice Now. Solve the triangle $\triangle ABC$ with $a = 6$, $b = 7$, and $A = 41°$. Round your answers to the nearest tenth.

Exercises 4.1

(1–4) Solve each triangle. Round solutions to the nearest tenth.

1.

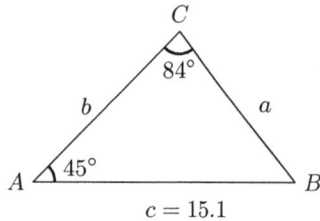

2.

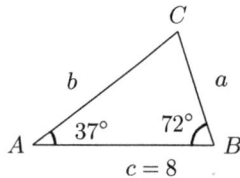

3.

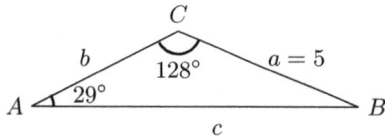

4.

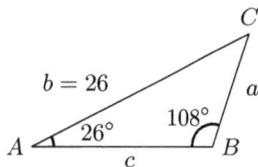

(5–10) Solve each triangle. Round the solutions to the nearest tenth.

5. $A = 47°$, $B = 27°$, $a = 25$

6. $A = 35°$, $C = 78°$, $b = 15$

7. $B = 15°$, $C = 25°$, $a = 7$

8. $B = 46°$, $C = 34°$, $b = 8$

9. $A = 105°$, $B = 16°$, $b = 14$

10. $A = 38°$, $C = 27°$, $c = 16$

(11–18) Determine the number of triangles that the given measurements produce. Then solve each triangle.

11. $A = 35°$, $a = 5$, $b = 15$

12. $A = 30°$, $a = 7$, $b = 14$

13. $A = 46°$, $a = 14$, $b = 12$

14. $A = 27°$, $a = 23$, $b = 30$

15. $B = 36°$, $b = 10$, $c = 14$

16. $B = 40°$, $b = 6$, $c = 10$

17. $C = 25°$, $c = 3$, $a = 4$

18. $C = 57°$, $c = 15$, $a = 17$

4.2 The Law of Cosines

> ✏️ **The Law of Cosines**
>
> If A, B, and C are three angles and a, b, and c are three sides of a triangle opposite to the corresponding angle, then
>
> $$a^2 = b^2 + c^2 - 2bc \cos A, \qquad \cos A = \frac{b^2 + c^2 - a^2}{2bc},$$
> $$b^2 = a^2 + c^2 - 2ac \cos B, \qquad \cos B = \frac{c^2 + a^2 - b^2}{2ca},$$
> $$c^2 = a^2 + b^2 - 2ab \cos C, \qquad \cos C = \frac{a^2 + b^2 - c^2}{2ab}.$$
>
> The square of any side of a triangle equals the sum of the squares of two other sides minus their product times the cosine of their included angle. Solve the law on the left for cosines to get the law on the right.

We will learn to use the law of cosines to solve two types of triangles—SSS and SAS triangles.

Example 1. Solving an SSS Triangle

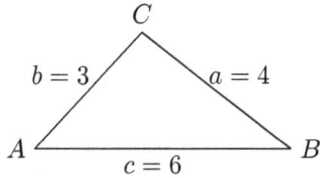

Solve the triangle $\triangle ABC$ with $a = 4$, $b = 3$, and $c = 6$.
Round your answers to the nearest tenth.

Solution. Because c is the longest side, we first use the law of cosines to find angle C.

$$\cos C = \frac{a^2 + b^2 - c^2}{2ab} = \frac{4^2 + 3^2 - 6^2}{2 \cdot 4 \cdot 3} \approx -0.45833$$

Then

$$C \approx \cos^{-1} -0.45833 \approx 117.3°$$

Next, use the law of sines to find angle B.

$$\sin B = \frac{b \sin C}{c} = \frac{3 \cdot \sin 117.3°}{6} \approx 0.44431$$

$$B \approx \sin^{-1} 0.44431 \approx 26.4°$$

Find the angle A by using the fact that $A + B + C = 180°$.

$$A = 180° - B - C \approx 180° - 26.4° - 117.3° = 36.3°.$$

Practice Now. Solve the triangle $\triangle ABC$ with $a = 5$, $b = 6$, and $c = 8$.

Example 2. Solving an SAS Triangle

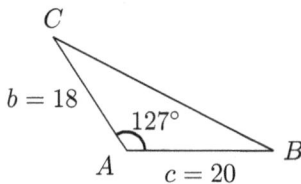

Solve the triangle $\triangle ABC$ with $b = 18$, $c = 20$, and $A = 127°$.
Round your answers to the nearest tenth.

Solution. Use the law of cosines to find side a.

$$a^2 = b^2 + c^2 - 2bc \cos A$$
$$= 18^2 + 20^2 - 2(18)(20) \cos 127°$$
$$\approx 1157.3$$
$$a \approx \sqrt{1157.3} \approx 34.0$$

Then we can use either the law of sines or the law of cosines for another angle. We demonstrate how to use the law of sines to find angle B.

$$\frac{\sin B}{b} = \frac{\sin A}{a}$$

$$\frac{\sin B}{18} = \frac{\sin 127°}{34}$$

$$\sin B = \frac{18 \sin 127°}{34} \approx 0.4228$$

This is an ambiguous case, so either $B = \sin^{-1} 0.4228 \approx 25.0°$ or $B \approx 180° - 25.0° = 155.0°$. But B should be an acute angle because A is already obtuse. Therefore, $B \approx 25.0°$. Finally,

$$C = 180° - A - B \approx 180° - 127° - 25.0° = 28.0°.$$

Practice Now.   Solve the triangle $\triangle ABC$ with $a = 12$, $b = 21$, and $C = 34°$. Round your answers to the nearest tenth.

▶ Area of a Triangle

We are all familiar with the area K of a triangle, which is $K = \frac{1}{2}(\text{base})(\text{height}) = \frac{1}{2}bh$, where b is the base and h is the height.

Using the definition of $\sin A$ on the left half of the triangle, we have

$$\sin A = \frac{h}{c}.$$

By multiplying both sides by c, we get $h = c \sin A$. Substituting this new expression for the height, we can write the area of an SAS triangle as

$$K = \frac{1}{2}bc \sin A.$$

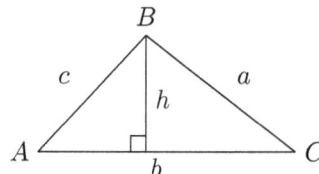

Simliar consideration with different orientations of the triangle gives us the following formulas.

✏ Area of an SAS Triangle

The area K of a triangle ABC with sides a, b, and c is

$$K = \frac{1}{2}bc \sin A \qquad K = \frac{1}{2}ca \sin B \qquad K = \frac{1}{2}ab \sin C$$

Example 3. Finding the Area of an SAS Triangle

Given the triangle on the right, find its area. Express the area rounded to three decimal places.

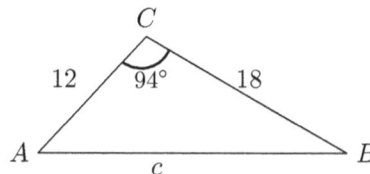

Solution.   Because the triangle is SAS, we can apply the formula for the area.

$$K = \frac{1}{2}ab \sin C = \frac{1}{2} \cdot 18 \cdot 12 \cdot \sin 94° = 107.737.$$

Practice Now.   In $\triangle ABC$, $b = 10$, $c = 8$, and the angle $A = 45°$. Find the area of $\triangle ABC$.

Example 4. Finding an Angle Given the Area of a Triangle

In $\triangle ABC$, $b = 12$ meters and $c = 20$ meters. If the area of the triangle is 77 square meters, find the measure of the angle A to the nearest degree.

Solution.   The values of b, c, and K are given, so we use the formula $K = \dfrac{1}{2}bc\sin A$. We substitute the known values and get the equation

$$77 = \frac{1}{2} \cdot 12 \cdot 20 \cdot \sin A.$$

Solve the equation for $\sin A$ to get $\sin A = \frac{77}{120}$. Now find solutions between $0°$ and $180°$.

$$A = \sin^{-1}\frac{77}{120} \approx 40° \quad \text{or} \quad 180° - 40° = 140°.$$

Two triangles with the given measures have the same area; one triangle is acute and the other is obtuse.

Practice Now.   In $\triangle ABC$, $a = 21$ inches and $b = 14$ inches. If the area of the triangle is 100 square inches, find the measure of the angle C. Round to the nearest degree.

Exercises 4.2

(1–4) Solve each triangle. Round solutions to the nearest tenth.

1.

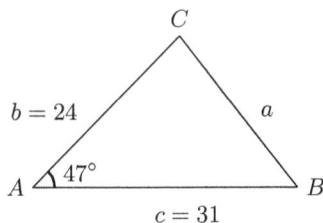

2.

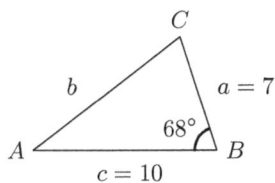

3.

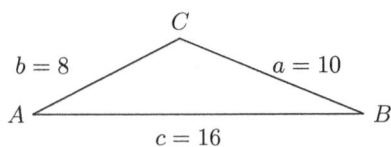

4.

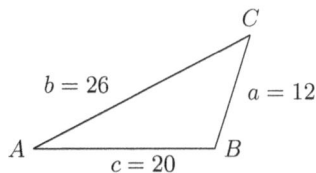

(5–10) Solve each triangle. Round solutions to the nearest tenth.

5. $a = 4$, $b = 7$, $C = 27°$

6. $b = 12$, $c = 18$, $A = 36°$

7. $a = 6$, $c = 9$, $B = 75°$

8. $a = 6$, $b = 3$, $C = 47°$

9. $b = 10$, $c = 16$, $A = 102°$

10. $a = 8$, $b = 3\sqrt{2}$, $C = 45°$

(11–18) Solve each triangle. Round solutions to the nearest tenth.

11. $a = 5$, $b = 6$, $c = 7$

12. $a = 12$, $b = 7$, $c = 9$

13. $a = 6$, $b = 3$, $c = 5$

14. $a = 27$, $b = 21$, $c = 30$

15. $a = 15$, $b = 10$, $c = 14$

16. $a = 9$, $b = 6$, $c = 12$

17. $a = 8$, $b = 3$, $c = 7$

18. $a = 5$, $b = 9$, $c = 13$

(19–21) Find the area of the triangle $\triangle ABC$.

19. $a = 24$, $b = 24$, $C = 30°$

20. $a = 6$, $c = 8$, $B = 87°$

21. $b = 3$, $c = 8$, $A = 98°$

4.3   Vectors

▶ The Definition of Vectors

A **vector** is a quantity with magnitude and direction. Vectors are used when we need both magnitude and direction, notably in physics for displacement, velocity, force, and so on.

We draw a vector geometrically as an arrow. The length of the arrow represents the magnitude, and the arrow tip indicates the direction. The name of a vector is written either in a bold letter (in print) or with a little arrow above the letter (in handwriting). Two vectors are defined to be equal if they have the same magnitude and direction regardless of their positions. The zero vector **0** is a special vector with no direction and zero magnitude.

Vectors drawn in the coordinate plane can be represented algebraically in numbers in the form of $\langle a, b \rangle$ (in angle brackets) where a is the horizontal component and b is the vertical component.

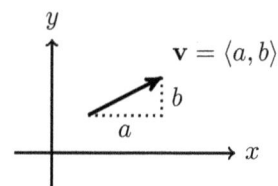

═══

Example 1. Drawing Vectors in the Coordinate Plane

Draw each vector in the coordinate plane.

a) $\mathbf{v} = \langle 3, -2 \rangle$ **b)** $\mathbf{w} = \langle 0, 1 \rangle$ **c)** $\mathbf{u} = \langle -2, -2 \rangle$

═══

Solution.   The horizontal componet of **v** is 3 and the vertical component is -2. Draw the vector in the direction of 3 units to the right and 2 units down. Other vectors are drawn similarly.

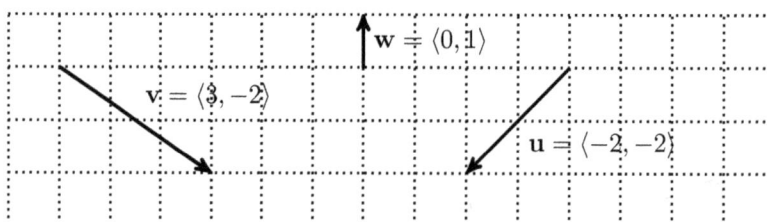

Practice Now.   Draw each vector in the coordinate plane.

a) $\mathbf{v} = \langle -2, 4 \rangle$ **b)** $\mathbf{w} = \langle 5, 0 \rangle$ **c)** $\mathbf{u} = \langle 1, -3 \rangle$

▶ Vector Operations—Addition and Scalar Multiplication

Vector Addition

Geometrically, we add vectors with the parallelogram law as follows. Draw a parallelogram using the given vectors. Draw an arrow diagonally from the common initial point. It is the sum of two vectors.

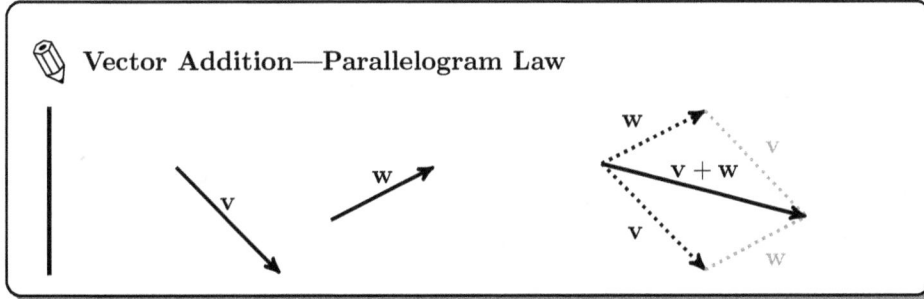

Vector Addition—Parallelogram Law

Algebraically, we add vectors componentwise. In other words, first components are added together, then second components are added together. If $\mathbf{v} = \langle v_1, v_2 \rangle$ and $\mathbf{w} = \langle w_1, w_2 \rangle$, then $\mathbf{v} + \mathbf{w} = \langle v_1 + w_1, v_2 + w_2 \rangle$.

Example 2. Adding Vectors Geometrically

Given vectors as follows, draw the following vectors: $\mathbf{v} + \mathbf{w}$, $\mathbf{v} + \mathbf{u}$, and $\mathbf{w} + \mathbf{u}$.

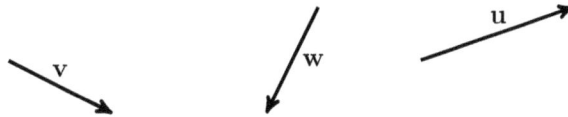

Solution. Apply the parallelogram law.

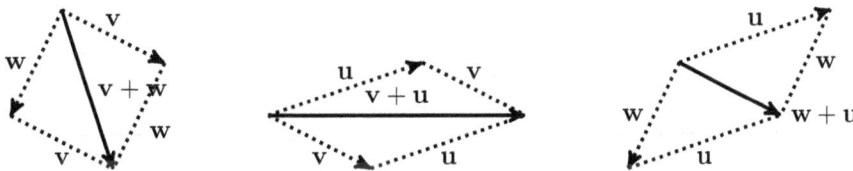

Practice Now. Draw the following vectors: $\mathbf{v} + \mathbf{w}$, $\mathbf{v} + \mathbf{u}$, and $\mathbf{w} + \mathbf{u}$.

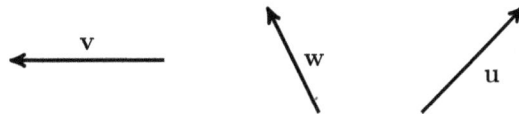

Example 3. Adding Vectors Algebraically

Let $\mathbf{v} = \langle 3, -2 \rangle$ and $\mathbf{w} = \langle 5, 6 \rangle$. Find the sum $\mathbf{v} + \mathbf{w}$.

Solution.   Add the given vectors componentwise.

$$\mathbf{v} + \mathbf{w} = \langle 3 + 5, -2 + 6 \rangle = \langle 8, 4 \rangle.$$

Practice Now.   Find the sum $\mathbf{v} + \mathbf{w}$.

a) $\mathbf{v} = \langle 5, -3 \rangle, \quad \mathbf{w} = \langle 2, 7 \rangle$ **b)** $\mathbf{v} = \langle 7, 10 \rangle, \quad \mathbf{w} = \langle -3, -5 \rangle$

Scalar Multiplication

When we multiply a vector $\mathbf{v}$ by a constant k, we get a new vector, written as $k\mathbf{v}$.

Geometrically, it is the vector scaled by the factor of $|k|$, keeping the same direction if $k > 0$, and reversing the direction if $k < 0$. If $k = 0$, we get the zero vector $\mathbf{0}$. The constant k is called a **scalar** because it scales vectors.

Algebraically, we multiply each component by the scalar.

Example 4. Multiplying Vectors by Scalars Geometrically

For the vector $\mathbf{v}$ on the right, draw each vector multiplied by a scalar.

a) $2\mathbf{v}$ **b)** $-\mathbf{v}$ **c)** $-\dfrac{\mathbf{v}}{2}$

Solution.   For $2\mathbf{v}$, draw the vector of twice the length, keeping the direction. For $-\mathbf{v}$, which is a shorthand for $(-1)\mathbf{v}$, reverse the direction, keeping the magnitude. For $-\dfrac{\mathbf{v}}{2}$, which is a shorthand for $\left(-\dfrac{1}{2}\right)\mathbf{v}$, reverse the direction, and reduce the length by half.

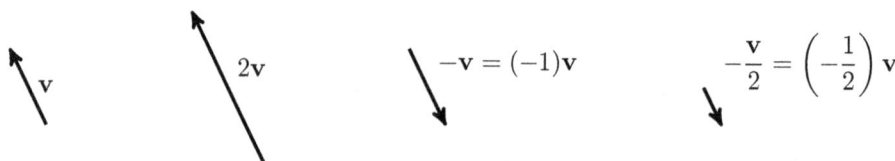

Practice Now.   For the given vector $\mathbf{v}$, draw each vector multiplied by a scalar.

a) $2\mathbf{v}$ **b)** $-\mathbf{v}$ **c)** $-\dfrac{\mathbf{v}}{2}$

Example 5. Multiplying Vectors by Scalars Algebraically

Let $\mathbf{v} = \langle 4, -5 \rangle$. Find each vector.

a) $3\mathbf{v}$ **b)** $-\mathbf{v}$ **c)** $\dfrac{\mathbf{v}}{2}$

Solution.

a) Multiply each component by 3.

$$3\mathbf{v} = \langle 3(4), 3(-5) \rangle = \langle 12, -15 \rangle$$

b) Negate each component.

$$-\mathbf{v} = \langle -(4), -(-5) \rangle = \langle -4, 5 \rangle$$

c) Divide each component by 2.

$$\frac{\mathbf{v}}{2} = \left\langle \frac{4}{2}, \frac{-5}{2} \right\rangle = \left\langle 2, -\frac{5}{2} \right\rangle$$

Practice Now. Let $\mathbf{v} = \langle 9, 10 \rangle$. Find each vector.

a) $5\mathbf{v}$ **b)** $-\mathbf{v}$ **c)** $-\dfrac{\mathbf{v}}{2}$

▶ **Position Vector**

> **Vector Components Given Initial and Terminal Points**
>
> The vector with the initial point P and the terminal point Q is written as $\overrightarrow{PQ}$. Its components are differences of coordinates.
>
> $P(x_1, y_1)$
>
> $Q(x_2, y_2)$
>
> $$\overrightarrow{PQ} = \langle x_2 - x_1, y_2 - y_1 \rangle$$

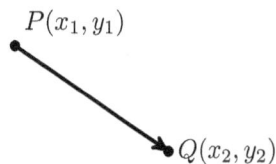

Example 6. Finding Vectors with Initial and Terminal Points

Draw the vector $\overrightarrow{PQ}$ in the coordinate plane and write it in components.

$$P(3, -1), \qquad Q(-2, 4)$$

Solution.

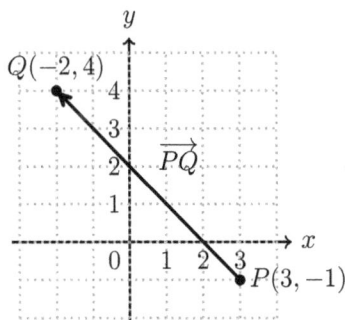

Plot the points P and Q in the coordinate plane, then draw the arrow from P to Q.
Use the difference formula to find the components.

$$\begin{aligned}\overrightarrow{PQ} &= \langle x_2 - x_1, y_2 - y_1 \rangle \\ &= \langle -2 - 3, 4 - (-1) \rangle \\ &= \langle -5, 5 \rangle\end{aligned}$$

Practice Now. Draw the vector $\overrightarrow{PQ}$ in the coordinate plane and write it in components.

a) $P(-1, 0), \quad Q(4, 1)$ **b)** $P(1, 3), \quad Q(2, -2)$

The **position vector** of a point is the vector with the initial point at the origin and the terminal point at the point.

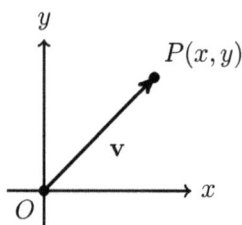

The components of the position vector of a point are the same as the coordinates of the point.

$$\mathbf{v} = \overrightarrow{OP} = \langle x - 0, y - 0 \rangle = \langle x, y \rangle$$

Position vectors encapsulate the coordinates of points in the form of a vector. We use position vectors because vectors are more flexible to work with than points. For example, we can add vectors, but we can't add points.

Example 7. Conversion between Points and Position Vectors

a) Find the position vector of $P(3, -4)$.

b) Find the point Q of which the position vector is $\langle -7, 5 \rangle$.

Solution. Remember to use correct notations. We use parentheses for coordinates of a point and angle brackets for components of a vector. Other than that, the answers are straightforward.

a) $\overrightarrow{OP} = \langle 3, -4 \rangle$.

b) $Q(-7, 5)$.

Practice Now.

a) Find the position vector of $P(7, 4)$.

b) Find the point Q of which the position vector is $\langle -2, 0 \rangle$.

▶ Magnitude and Unit Vectors

Using the Pythagorean theorem, we get the following magnitude formula.

> **✎ The Magnitude of a Vector**
>
> For a vector $\mathbf{v} = \langle a, b \rangle$, its magnitude is denoted by $\|\mathbf{v}\|$ and
>
> $$\|\mathbf{v}\| = \sqrt{a^2 + b^2}.$$
>
>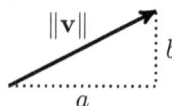

Example 8. Finding the Magnitude of a Vector

Find the magnitude of each vector.

a) $\mathbf{v} = \langle -3, 3 \rangle$

b) $\mathbf{w} = \langle 0, -4 \rangle$

Solution. Apply the magnitude formula.

a) $\|\mathbf{v}\| = \sqrt{(-3)^2 + 3^2} = \sqrt{9+9} = \sqrt{18} = \sqrt{2 \cdot 3^2} = 3\sqrt{2}.$

b) $\|\mathbf{w}\| = \sqrt{0^2 + (-4)^2} = \sqrt{(-4)^2} = \sqrt{16} = 4.$

Practice Now. Find the magnitude of each vector.

  a) $\mathbf{v} = \langle -3, -4 \rangle$ **b)** $\mathbf{w} = \langle -7, 0 \rangle$

A vector of magnitude 1 is called a **unit vector**. We get a unit vector in the direction of another vector by dividing it by its length, normalizing the magnitude and keeping the direction. For example, if $\|\mathbf{v}\| = 3$, then $\dfrac{\mathbf{v}}{3}$ is a vector of magnitude 1 that points in the same direction.

$\|\mathbf{v}\| = 3$ $\dfrac{\mathbf{v}}{3}$: unit vector

✏️ **Unit Vector Formula**

A unit vector $\mathbf{u}$ in the direction of a vector $\mathbf{v}$ is

$$\mathbf{u} = \frac{\mathbf{v}}{\|\mathbf{v}\|}.$$

Example 9. Finding a Unit Vector in the Direction of a Vector

Find a unit vector $\mathbf{u}$ in the direction of $\mathbf{v} = \langle -2, 4 \rangle$.

Solution. Use the magnitude formula to get $\|\mathbf{v}\| = \sqrt{(-2)^2 + 4^2} = \sqrt{20} = 2\sqrt{5}.$ Then use the unit vector formula.

$$\begin{aligned}
\mathbf{u} &= \frac{\mathbf{v}}{\|\mathbf{v}\|} = \frac{\langle -2, 4 \rangle}{2\sqrt{5}} \\[2mm]
&= \left\langle \frac{-2}{2\sqrt{5}}, \frac{4}{2\sqrt{5}} \right\rangle = \left\langle -\frac{1}{\sqrt{5}}, \frac{2}{\sqrt{5}} \right\rangle \qquad \text{(Divide and simplify.)} \\[2mm]
&= \left\langle -\frac{\sqrt{5}}{5}, \frac{2\sqrt{5}}{5} \right\rangle \qquad\qquad\qquad \text{(Rationalize the denominator.)}
\end{aligned}$$

Practice Now. Find a unit vector in the direction of each vector.

  a) $\mathbf{v} = \langle 8, -6 \rangle$ **b)** $\mathbf{w} = \langle 3, 1 \rangle$

▶ **Standard Unit Vectors**

Standard unit vectors are unit vectors in the direction of coordinate axes. They are named $\mathbf{i}$ and $\mathbf{j}$.

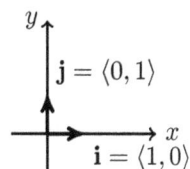

$\mathbf{j} = \langle 0, 1 \rangle$

$\mathbf{i} = \langle 1, 0 \rangle$

Every vector can be written as a "linear combination" of standard unit vectors and vice versa.

$$\langle a, b \rangle = a\mathbf{i} + b\mathbf{j}$$

For example, the vector $\mathbf{v} = \langle 4, 3 \rangle$ can be written in terms of standard unit vectors as $\mathbf{v} = 4\mathbf{i} + 3\mathbf{j}$. Or, the vector $\mathbf{w} = -6\mathbf{i} + 7\mathbf{j}$ is equal to $\langle -6, 7 \rangle$.

▶ Direction Angle

Suppose a vector $\mathbf{v} = \langle a, b \rangle$ is drawn as a position vector with the initial point at the origin. The **direction angle** θ of $\mathbf{v}$ is the smallest non-negative angle from the positive x-axis to the vector in standard position. A direction angle together with magnitude characterizes a vector. The conversion formula from magnitude and direction angle to components is as follows.

✎ Vector Components Given Magnitude and Direction Angle

If θ is the direction angle of a vector $\mathbf{v}$, then

$$\mathbf{v} = \|\mathbf{v}\| \langle \cos\theta, \sin\theta \rangle$$

Example 10. Finding Vector Components with Magnitude and Direction Angle

Find the vector $\mathbf{v}$ with magnitude 6 and direction angle $135°$.

Solution. Apply the formula, find trigonometric values, then simplify numbers.

$$\mathbf{v} = 6 \langle \cos 135°, \sin 135° \rangle$$
$$= \langle 6\cos 135°, 6\sin 135° \rangle$$
$$= \left\langle 6\left(-\frac{\sqrt{2}}{2}\right), 6\left(\frac{\sqrt{2}}{2}\right) \right\rangle$$
$$= \left\langle -3\sqrt{2}, 3\sqrt{2} \right\rangle.$$

Practice Now. Find the vector with the given magnitude and direction angle.

a) $\|\mathbf{v}\| = 8, \quad \theta = 120°$
b) $\|\mathbf{v}\| = 1, \quad \theta = 300°$

Conversely, we can find the direction angle using components. If the vector $\mathbf{v}$ belongs to one of four quadrants, then the direction angle θ can be obtained by solving a tangent equation.

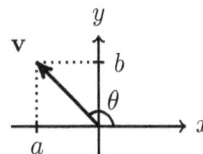

Direction Angle

If $\mathbf{v} = \langle a, b \rangle$ and $a \neq 0$, then its direction angle θ satisfies

$$\tan\theta = \frac{b}{a}.$$

Example 11. Finding the Direction Angle of a Vector

Find the direction angle of each vector.

a) $\mathbf{v} = \langle -3, -2 \rangle$ b) $\mathbf{v} = \langle 4, 4 \rangle$ c) $\mathbf{v} = \langle 1, -2 \rangle$ d) $\mathbf{v} = \langle 0, -2 \rangle$

Solution.

a) The vector is in quadrant III. Because $\tan^{-1}\left(\frac{-2}{-3}\right) = 33.7°$ is an angle of quadrant I, adjust it by adding $180°$. The answer is $213.7°$.

b) The vector is in quadrant I. Because $\tan^{-1}\left(\frac{4}{4}\right) = 45°$ is already in quadrant I, the answer is $45°$.

c) The vector is in quadrant IV. Because $\tan^{-1}\left(\frac{-2}{1}\right) = -63.4°$ is in quadrant IV but is of negative value, adjust it by adding $360°$. The answer is $296.6°$.

d) The vector points in the negative y direction. The angle is $270°$. (We don't even need to use the inverse tangent function. In fact, $y/x = -2/0$ is undefined.)

Practice Now.   Find the direction angle of each vector.

a) $\mathbf{v} = \langle 3, 4 \rangle$ b) $\mathbf{v} = \langle -3, -4 \rangle$ c) $\mathbf{v} = \langle -7, 0 \rangle$ d) $\mathbf{v} = \langle 0, 5 \rangle$

Exercises 4.3

1. Write each vector in components.

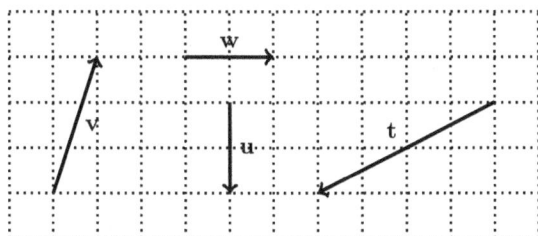

(2–4) Draw the vectors in the coordinate plane.

2. $\mathbf{v} = \langle 6, 1 \rangle$, $\mathbf{w} = \langle 2, -3 \rangle$

3. $\mathbf{v} = \langle -3, 4 \rangle$, $\mathbf{w} = \langle -1, -1 \rangle$

4. $\mathbf{v} = \langle 0, -3 \rangle$, $\mathbf{w} = \langle 3, 0 \rangle$

(5–10) Use the vectors in the previous problem to draw each of the following vectors.

5. $-\mathbf{v}$ 6. $\mathbf{v} + \mathbf{w}$

7. $\frac{1}{2}\mathbf{t}$ 8. $2\mathbf{w} - 3\mathbf{u}$

9. $\mathbf{t} - \mathbf{v}$ 10. $\mathbf{v} + \mathbf{w} + \mathbf{u}$

(11–16) Let $\mathbf{v} = \langle 5, 0 \rangle$, $\mathbf{w} = \langle -1, 3 \rangle$, and $\mathbf{u} = \langle 2, -4 \rangle$. Find each of the following vectors.

11. $\mathbf{v} + \mathbf{w}$ 12. $\mathbf{v} - \mathbf{u}$

13. $5\mathbf{w} + 4\mathbf{u}$

14. $-3\mathbf{w} + \dfrac{1}{5}\mathbf{v}$

15. $\mathbf{v} + \mathbf{w} - \mathbf{u}$

16. $\dfrac{\mathbf{u}}{2} - 3\mathbf{v} + \mathbf{w}$

(17–22) Suppose three points $P(7,9)$, $Q(-3,4)$, and $R(2,-2)$ are given. Find each of the following vectors.

17. $\overrightarrow{PQ}$

18. $\overrightarrow{QP}$

19. $\overrightarrow{QR}$

20. $\overrightarrow{RQ}$

21. $\overrightarrow{RP}$

22. $\overrightarrow{PR}$

(23–26) Write the position vector of the point, and draw it in the xy-plane.

23. $P(2,2)$

24. $Q(-3,0)$

25. $R(5,-1)$

26. $S(-2,-3)$

(27–32) Find the magnitude $\|\mathbf{v}\|$ and the direction angle θ of the vector $\mathbf{v}$.

27. $\mathbf{v} = \langle 2,6 \rangle$

28. $\mathbf{v} = \langle -1,1 \rangle$

29. $\mathbf{v} = \langle 2\sqrt{3}, -2 \rangle$

30. $\mathbf{v} = \langle -12, -5 \rangle$

31. $\mathbf{v} = \langle -3, 0 \rangle$

32. $\mathbf{v} = \langle 0, 5 \rangle$

(33–38) Find a unit vector $\mathbf{u}$ in the direction of the given vector.

33. $\mathbf{v} = \langle -4, -3 \rangle$

34. $\mathbf{v} = \langle -5, 12 \rangle$

35. $\mathbf{v} = \langle -2, 2 \rangle$

36. $\mathbf{v} = \langle 3, 6 \rangle$

37. $\mathbf{v} = \langle 0, -6 \rangle$

38. $\mathbf{v} = \langle 5, 0 \rangle$

(39–44) Find the components of the vector $\mathbf{v}$ with the given magnitude and direction angle θ.

39. $\|\mathbf{v}\| = 6$, $\quad \theta = 30°$

40. $\|\mathbf{v}\| = 4$, $\quad \theta = 120°$

41. $\|\mathbf{v}\| = 5$, $\quad \theta = 225°$

42. $\|\mathbf{v}\| = 7$, $\quad \theta = 330°$

43. $\|\mathbf{v}\| = 1$, $\quad \theta = 90°$

44. $\|\mathbf{v}\| = 4$, $\quad \theta = 0°$

4.4 The Dot Product of Vectors

In this section we discuss the dot product of two vectors. It is obtained by multiplying vectors with the same number of components, and results in a number.

▶ **The Dot Product**

✎ **The Dot Product**

The **dot product of two vectors** $\mathbf{u} = \langle a, b \rangle$ and $\mathbf{v} = \langle c, d \rangle$ is defined by

$$\mathbf{u} \cdot \mathbf{v} = ac + bd$$

Example 1. Finding the Dot Product using Definition

Find the dot product $\mathbf{u} \cdot \mathbf{v}$.

a) $\mathbf{u} = \langle 1, 3 \rangle$, $\quad \mathbf{v} = \langle 4, -2 \rangle$

b) $\mathbf{u} = \mathbf{i} + 4\mathbf{j}$, $\quad \mathbf{v} = 8\mathbf{i} - \mathbf{j}$

Solution.

a) $\mathbf{u} \cdot \mathbf{v} = \langle 1, 3 \rangle \cdot \langle 4, -2 \rangle = (1)(4) + (3)(-2) = -2$

b) $\mathbf{u} \cdot \mathbf{v} = (\mathbf{i} + 4\mathbf{j}) \cdot (8\mathbf{i} - \mathbf{j}) = \langle 1, 4 \rangle \cdot \langle 8, -1 \rangle = (1)(8) + (4)(-1) = 4$

Practice Now. Find the dot product $\mathbf{u} \cdot \mathbf{v}$.

a) $\mathbf{u} = \langle 2, -11 \rangle$, $\quad \mathbf{v} = \langle -8, 7 \rangle$

b) $\mathbf{u} = 6\mathbf{i} - 2\mathbf{j}$, $\quad \mathbf{v} = 8\mathbf{i} + \mathbf{j}$

▶ Properties of the Dot Product

If $\mathbf{u}, \mathbf{v}$, and $\mathbf{w}$ are vectors and k is a scalar, then

- $\mathbf{u} \cdot \mathbf{v} = \mathbf{v} \cdot \mathbf{u}$
- $\mathbf{u} \cdot \mathbf{u} = \|\mathbf{u}\|^2$
- $\mathbf{0} \cdot \mathbf{u} = 0$
- $k(\mathbf{u} \cdot \mathbf{v}) = (k\mathbf{u}) \cdot \mathbf{v} = \mathbf{u} \cdot (k\mathbf{v})$
- $(\mathbf{u} + \mathbf{v}) \cdot \mathbf{w} = \mathbf{u} \cdot \mathbf{w} + \mathbf{v} \cdot \mathbf{w}$

- $\mathbf{u} \cdot (\mathbf{v} + \mathbf{w}) = \mathbf{u} \cdot \mathbf{v} + \mathbf{u} \cdot \mathbf{w}$
- $\|\mathbf{u} + \mathbf{v}\|^2 = \|\mathbf{u}\|^2 + \|\mathbf{v}\|^2 + 2(\mathbf{u} \cdot \mathbf{v})$
- $\|\mathbf{u} - \mathbf{v}\|^2 = \|\mathbf{u}\|^2 + \|\mathbf{v}\|^2 - 2(\mathbf{u} \cdot \mathbf{v})$
- $(\mathbf{u} + \mathbf{v}) \cdot (\mathbf{u} - \mathbf{v}) = \|\mathbf{u}\|^2 - \|\mathbf{v}\|^2$

▶ The Angle between Two Vectors

The Dot Product and the Angle between Two Vectors

If θ is the angle between two nonzero vectors $\mathbf{u} = \langle a, b \rangle$ and $\mathbf{v} = \langle c, d \rangle$, where $0° \leq \theta \leq 180°$, then

$$\mathbf{u} \cdot \mathbf{v} = \|\mathbf{u}\|\|\mathbf{v}\| \cos\theta,$$

$$\cos\theta = \frac{\mathbf{u} \cdot \mathbf{v}}{\|\mathbf{u}\|\|\mathbf{v}\|}.$$

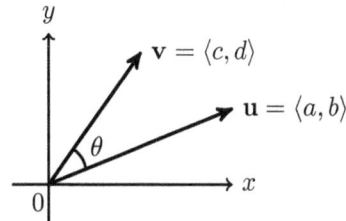

Example 2. Finding the Dot Product with Magnitudes and the Angle between Vectors

Find $\mathbf{u} \cdot \mathbf{v}$, where θ is the angle between the vectors $\mathbf{u}$ and $\mathbf{v}$. Round the answer to the nearest tenth.

a) $\|\mathbf{u}\| = 2, \ \|\mathbf{v}\| = 6, \ \theta = \dfrac{\pi}{4}$

b) $\|\mathbf{u}\| = 3, \ \|\mathbf{v}\| = 8, \ \theta = 25°$

Solution.

a) $\mathbf{u} \cdot \mathbf{v} = \|\mathbf{u}\|\|\mathbf{v}\| \cos\theta = (2)(6) \cos\dfrac{\pi}{4} = \dfrac{12\sqrt{2}}{2} = 6\sqrt{2} \approx 8.5$

b) $\mathbf{u} \cdot \mathbf{v} = \|\mathbf{u}\|\|\mathbf{v}\| \cos\theta = (3)(8) \cos 25° = 24 \cos 25° \approx 21.8$

Practice Now. Find $\mathbf{u} \cdot \mathbf{v}$, where θ is the angle between $\mathbf{u}$ and $\mathbf{v}$. Round the answer to the nearest tenth.

a) $\|\mathbf{u}\| = 7, \ \|\mathbf{v}\| = 8, \ \theta = \dfrac{\pi}{3}$

b) $\|\mathbf{u}\| = 2, \ \|\mathbf{v}\| = 5, \ \theta = 127°$

Example 3. Finding the Angle between Vectors

Find the angle θ between the vectors $\mathbf{u}$ and $\mathbf{v}$. Round the answer to the nearest tenth of a degree.

a) $\mathbf{u} = \langle 2, 4 \rangle, \ \mathbf{v} = \langle 6, 3 \rangle$

b) $\mathbf{u} = \mathbf{i} - \mathbf{j}, \ \mathbf{v} = 2\mathbf{i} + 3\mathbf{j}$

Solution.

a) Step 1. $\mathbf{u} \cdot \mathbf{v} = \langle 2, 4 \rangle \cdot \langle 6, 3 \rangle = (2)(6) + (4)(3) = 24.$

Step 2. $\|\mathbf{u}\| = \sqrt{\mathbf{u} \cdot \mathbf{u}} = \sqrt{2^2 + 4^2} = \sqrt{20}$ and $\|\mathbf{v}\| = \sqrt{\mathbf{v} \cdot \mathbf{v}} = \sqrt{6^2 + 3^2} = \sqrt{45}.$

Step 3. $\cos\theta = \dfrac{\mathbf{u} \cdot \mathbf{v}}{\|\mathbf{u}\|\|\mathbf{v}\|} = \dfrac{24}{\sqrt{20} \cdot \sqrt{45}}$

Step 4. $\theta = \cos^{-1}\left(\dfrac{\mathbf{u} \cdot \mathbf{v}}{\|\mathbf{u}\|\|\mathbf{v}\|}\right) = \cos^{-1}\left(\dfrac{24}{\sqrt{20} \cdot \sqrt{45}}\right) \approx 36.9°$

b) Step 1. $\mathbf{u} \cdot \mathbf{v} = (\mathbf{i} - \mathbf{j}) \cdot (2\mathbf{i} + 3\mathbf{j}) = (1)(2) + (-1)(3) = -1.$

Step 2. $\|\mathbf{u}\| = \sqrt{\mathbf{u} \cdot \mathbf{u}} = \sqrt{1^2 + (-1)^2} = \sqrt{2}$ and $\|\mathbf{v}\| = \sqrt{\mathbf{v} \cdot \mathbf{v}} = \sqrt{2^2 + 3^2} = \sqrt{13}.$

Step 3. $\cos\theta = \dfrac{\mathbf{u} \cdot \mathbf{v}}{\|\mathbf{u}\|\|\mathbf{v}\|} = \dfrac{-1}{\sqrt{2} \cdot \sqrt{13}}$

Step 4. $\theta = \cos^{-1}\left(\dfrac{\mathbf{u} \cdot \mathbf{v}}{\|\mathbf{u}\|\|\mathbf{v}\|}\right) = \cos^{-1}\left(\dfrac{-1}{\sqrt{2} \cdot \sqrt{13}}\right) \approx 101.3°$

Practice Now.   Find the angle θ between $\mathbf{u}$ and $\mathbf{v}$. Round the answer to the nearest tenth of a degree.

a) $\mathbf{u} = \langle -1, 3 \rangle, \quad \mathbf{v} = \langle -4, 5 \rangle$ **b)** $\mathbf{u} = 2\mathbf{i} + 5\mathbf{j}, \quad \mathbf{v} = 4\mathbf{i} - 3\mathbf{j}$

Example 4. Finding the Magnitude of a Combination of Vectors

Let $\mathbf{u}$ and $\mathbf{v}$ be two vectors with $\|\mathbf{u}\| = 10$ and $\|\mathbf{v}\| = 15$. The angle between $\mathbf{u}$ and $\mathbf{v}$ is $43°$. Find $\|\mathbf{u} - 2\mathbf{v}\|$. Round the answer to the nearest tenth.

Solution.   It is easier to find the squared magnitude $\|\mathbf{u} - 2\mathbf{v}\|^2$. Expand it as follows.

$$\begin{aligned}
\|\mathbf{u} - 2\mathbf{v}\|^2 &= (\mathbf{u} - 2\mathbf{v}) \cdot (\mathbf{u} - 2\mathbf{v}) \\
&= \mathbf{u} \cdot \mathbf{u} - 2(\mathbf{u} \cdot \mathbf{v}) - 2(\mathbf{v} \cdot \mathbf{u}) + 4(\mathbf{v} \cdot \mathbf{v}) \\
&= \|\mathbf{u}\|^2 - 4(\mathbf{u} \cdot \mathbf{v}) + 4\|\mathbf{v}\|^2 \\
&= \|\mathbf{u}\|^2 - 4\|\mathbf{u}\|\|\mathbf{v}\|\cos\theta + 4\|\mathbf{v}\|^2 \\
&= (10)^2 - 4(10)(15)\cos 43° + 4(15)^2 \approx 561.188
\end{aligned}$$

Now take square roots of both sides for the solution $\|\mathbf{u} - 2\mathbf{v}\| \approx 23.7.$

Practice Now.   Let $\mathbf{u}$ and $\mathbf{v}$ be two vectors with $\|\mathbf{u}\| = 9$ and $\|\mathbf{v}\| = 5$. The angle between $\mathbf{u}$ and $\mathbf{v}$ is $29°$. Find $\|\mathbf{u} + 2\mathbf{v}\|$. Round the answer to the nearest tenth.

▶ Orthogonal Vectors

When two vectors $\mathbf{u}$ and $\mathbf{v}$ are **parallel**, the angle between them is $0°$ (same direction) or $180°$ (opposite direction. When two vectors $\mathbf{u}$ and $\mathbf{v}$ are **orthogonal (perpendicular)**, the angle between them is $\theta = 90°$ or $\frac{\pi}{2}$. Because $\cos 90° = 0$, the dot product is $\mathbf{u} \cdot \mathbf{v} = \|\mathbf{u}\|\|\mathbf{v}\|\cos 90° = 0$. Conversely, if $\mathbf{u} \cdot \mathbf{v} = 0$, then $\cos\theta = 0$, so $\theta = 90°$.

> ✏️ **Orthogonal Vectors**
>
> Two vectors $\mathbf{u}$ and $\mathbf{v}$ are **orthogonal (perpendicular)** if and only if
>
> $$\mathbf{u} \cdot \mathbf{v} = 0.$$

Example 5. Determining Orthogonality of Vectors

Determine whether each pair of vectors is orthogonal.

  a) $\mathbf{u} = \langle 3, 2 \rangle$ $\mathbf{v} = \langle -6, 9 \rangle$ | **b)** $\mathbf{u} = -7\mathbf{i} - 3\mathbf{j}$ $\mathbf{v} = 7\mathbf{i} + 3\mathbf{j}$

Solution.

  a) $\mathbf{u} \cdot \mathbf{v} = (3)(-6) + (2)(9) = 0$. Because $\mathbf{u} \cdot \mathbf{v} = 0$, vectors $\mathbf{u}$ and $\mathbf{v}$ are orthogonal.

  b) $\mathbf{u} \cdot \mathbf{v} = (-7\mathbf{i} - 3\mathbf{j}) \cdot (7\mathbf{i} + 3\mathbf{j}) = \langle -7, -3 \rangle \cdot \langle 7, 3 \rangle = (-7)(7) + (-3)(3) = -58$. Because $\mathbf{u} \cdot \mathbf{v} \neq 0$, vectors $\mathbf{u}$ and $\mathbf{v}$ are not orthogonal.

Practice Now.   Determine whether each pair of vectors is orthogonal.

  a) $\mathbf{u} = \langle -4, 3 \rangle$ $\mathbf{v} = \langle -5, -9 \rangle$ | **b)** $\mathbf{u} = 3\mathbf{i} - 4\mathbf{j}$ $\mathbf{v} = -8\mathbf{i} - 6\mathbf{j}$

Example 6. Finding an Orthogonal Vector

Find the scalar k so that the vectors $\mathbf{u} = \langle 1, 2 \rangle$ and $\mathbf{v} = \langle k, -5 \rangle$ are orthogonal.

Solution.   By the definition of orthogonality, $(1)(k) + (2)(-5) = 0$. $k + (-10) = 0$. Then $k = 10$.

Practice Now.   Find the scalar k so that the vectors $\mathbf{u} = \langle -40, 8 \rangle$ and $\mathbf{v} = \langle k, -2 \rangle$ are orthogonal.

▶ Work

For moving objects, the work W done by a constant force $\mathbf{F}$ on a point that moves a displacement (not distance) $\mathbf{d}$ in the direction of the force is defined by their product.

> ✏️ **Work**
>
> The work W done by a constant force $\mathbf{F}$ in moving an object from a point A to a point B is defined by
>
> $$W = \mathbf{F} \cdot \mathbf{d} = \|\mathbf{F}\| \|\mathbf{d}\| \cos \theta$$
>
> where $\mathbf{d}$ is the displacement vector and θ is the angle between $\mathbf{F}$ and $\mathbf{d}$.

Example 7. Finding the Work Done by a Force on an Object

How much work is done when a force (in pounds) $\mathbf{F} = \langle 4, 2 \rangle$ moves an object from $(0, 0)$ to $(6, 7)$ (the distance in feet)?

Solution. The displacement vector is $\mathbf{d} = \langle 6, 7 \rangle$. Then the work $W = \langle 4, 2 \rangle \cdot \langle 6, 7 \rangle = (4)(6) + (2)(7) = 38$ (ft-lb).

Practice Now. How much work is done when a force (in pounds) $\mathbf{F} = \langle 1, 3 \rangle$ moves an object from $(0, 0)$ to $(5, 2)$ (the distance in feet)?

Example 8. Finding the Work Done by Force (Application)

A wagon is pulled a distance of 10 feet along a horizontal path by a constance force of 60 pounds. The handle of wagon is held at an angle of $60°$ above the horizontal. Find the work done by the force.

Solution. $W = \|\mathbf{F}\| \|\mathbf{d}\| \cos \theta = (60)(10) \cos 60° = 300$ (ft-lb).

Practice Now. A sled is pulled along a level path through snow by a rope. A 20-lb force acting at an angle of $25°$ above the horizontal moves the sled 40 ft. Find the work done by the force.

Exercises 4.4

(1–10) Find the dot product $\mathbf{u} \cdot \mathbf{v}$.

1. $\mathbf{u} = \langle -2, 3 \rangle, \quad \mathbf{v} = \langle 5, 2 \rangle$

2. $\mathbf{u} = \langle -7, 4 \rangle, \quad \mathbf{v} = \langle 3, 2 \rangle$

3. $\mathbf{u} = 2\mathbf{i} + \mathbf{j}, \quad \mathbf{v} = \mathbf{i} - \mathbf{j}$

4. $\mathbf{u} = 3\mathbf{i} + 2\mathbf{j}, \quad \mathbf{v} = 4\mathbf{i} + 5\mathbf{j}$

5. $\mathbf{u} = \mathbf{i}, \quad \mathbf{v} = \mathbf{i}$

6. $\mathbf{u} = \mathbf{i}, \quad \mathbf{v} = \mathbf{j}$

7. $\|\mathbf{u}\| = 6, \quad \|\mathbf{v}\| = 5, \theta = \dfrac{\pi}{4}.$

8. $\|\mathbf{u}\| = 3, \quad \|\mathbf{v}\| = 4, \theta = \dfrac{\pi}{3}.$

9. $\|\mathbf{u}\| = 5, \quad \|\mathbf{v}\| = \sqrt{2}, \theta = 135°.$

10. $\|\mathbf{u}\| = 3, \quad \|\mathbf{v}\| = \sqrt{2}, \theta = 120°.$

(11–14) Find the angle between two vectors. Round the answer to the nearest tenth of degree.

11. $\mathbf{u} = \langle 1, -1 \rangle, \quad \mathbf{v} = \langle 5, 6 \rangle$

12. $\mathbf{u} = \langle -2, 5 \rangle, \quad \mathbf{v} = \langle 3, 2 \rangle$

13. $\mathbf{u} = 4\mathbf{i} - 3\mathbf{j}, \quad \mathbf{v} = 2\mathbf{i} - \mathbf{j}$

14. $\mathbf{u} = \mathbf{i} + 2\mathbf{j}, \quad \mathbf{v} = 4\mathbf{i} + 7\mathbf{j}$

(15–18) Let $\mathbf{u}$ and $\mathbf{v}$ be two vectors with given magnitudes. θ is the angle between $\mathbf{u}$ and $\mathbf{v}$. Find the indicated expression. Round the answer to the nearest hundredth.

15. $\|\mathbf{u}\| = 14, \|\mathbf{v}\| = 12, \theta = 60°$ Find $\|\mathbf{u} + \mathbf{v}\|$.

16. $\|\mathbf{u}\| = 6, \|\mathbf{v}\| = 10, \theta = 29°$ Find $\|\mathbf{u} + 3\mathbf{v}\|$.

17. $\|\mathbf{u}\| = 3, \|\mathbf{v}\| = 7, \theta = 27°$ Find $\|\mathbf{u} - \mathbf{v}\|$.

18. $\|\mathbf{u}\| = 12, \|\mathbf{v}\| = 7, \theta = 67°$ Find $\|\mathbf{u} - 2\mathbf{v}\|$.

(19–26) Determine whether each pair of vectors is orthogonal.

19. $\mathbf{u} = \langle 4, 6 \rangle, \quad \mathbf{v} = \langle -3, 2 \rangle$

20. $\mathbf{u} = \langle -2, -1 \rangle, \quad \mathbf{v} = \langle -3, 6 \rangle$

21. $\mathbf{u} = \langle -5, 3 \rangle, \quad \mathbf{v} = \langle 6, -8 \rangle$

22. $\mathbf{u} = \langle 2, 5 \rangle, \quad \mathbf{v} = \langle -3, 2 \rangle$

23. $\mathbf{u} = -\mathbf{i} + 2\mathbf{j}, \quad \mathbf{v} = 6\mathbf{i} + 3\mathbf{j}$

24. $\mathbf{u} = -6\mathbf{i} + 3\mathbf{j}, \quad \mathbf{v} = 2\mathbf{i} + 4\mathbf{j}$

25. $\mathbf{u} = -\mathbf{i} + 2\mathbf{j}, \quad \mathbf{v} = 3\mathbf{i} + 4\mathbf{j}$

26. $\mathbf{u} = 2\mathbf{i} + 6\mathbf{j}, \quad \mathbf{v} = -3\mathbf{i} - 9\mathbf{j}$

(27–29) Find the work done by force. Round the answer to the nearest tenth, if possible.

27. A force of 6 lbs acts in the direction of $10°$ to the horizontal. The force moves an object along a straight line from the origin to the point $(5, 12)$. Distance is measured in feet.

28. A constnce force $\mathbf{F} = 4\mathbf{i} + 13\mathbf{j}$ moves an object from the point $A(3, 4)$ to the point $B(7, 13)$, where units are in pounds and feet. Find the work done.

29. A person pulls a freight cart 90 feet with a force 34 pounds. How much work is done in moving the cart if the cart's handle makes an angle of $14°$ with the ground.

4.5 Polar Form of Complex Numbers

▶ The Complex Plane

The coordinate plane representing complex numbers is called the **complex plane**. The horizontal axis is called the **real axis**, and the vertical axis is called the **imaginary axis**. There is one-to-one correspondence between the complex numbers and the points on the complex plane. The correspondence is

$$a + bi \quad \longleftrightarrow \quad (a, b).$$

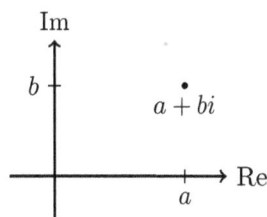

Example 1. Plotting Complex Numbers in the Complex Plane

Plot each complex number in the complex plane.

a) $2 - i$ **b)** $-2i$ **c)** -2 **d)** $-3 + i$

Solution. Convert each number into coordinates and plot it at the location.

a) $2 - i = 2 + (-1)i \quad \longrightarrow \quad (2, -1)$

b) $-2i = 0 + (-2)i \quad \longrightarrow \quad (0, -2)$

c) $-2 = (-2) + (0)i \quad \longrightarrow \quad (-2, 0)$

d) $-3 + i = -3 + (1)i \quad \longrightarrow \quad (-3, 1)$

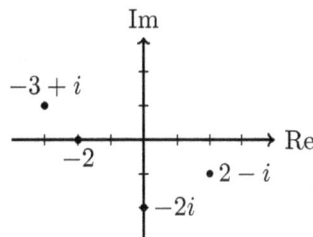

Practice Now. Plot each complex number in the complex plane.

a) $3 + 2i$ **b)** $4i$ **c)** 0 **d)** $1 - 2i$

▶ The Polar Form of a Complex Number

The **absolute value** (or **modulus**) $|z|$ of a complex number z is defined to be the distance from zero to the number in the complex plane. The **argument** θ of z (if $z \neq 0$) is defined to be the angle in standard position from the positive x-axis to the line from the origin to z.

By the Pythagorean theorem, if $z = a + bi$, then

$$|z| = \sqrt{a^2 + b^2},$$

and by definition of the tangent function, the argument θ satisfies the equation

$$\tan \theta = \frac{b}{a}.$$

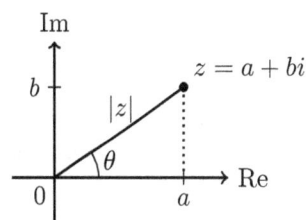

A complex number z written in the form of $a + bi$ is said to be in rectangular form because its position in the complex plane is found by drawing a rectangle. The position of z can be determined also in terms of modulus r and argument θ as $z = r(\cos\theta + i\sin\theta)$. The complex number written in this form is said to be in polar form because the position of z is described in reference to the fixed point at 0, which is called the **pole**.

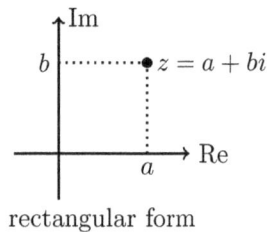

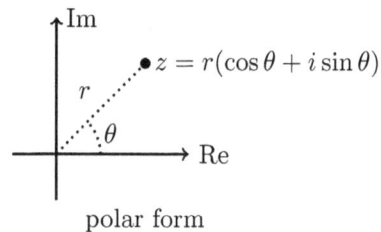

rectangular form   polar form

We convert between rectangular form and polar form with the following formulas, which are similar to the conversion formulas for vectors.

Conversion between Rectangular and Polar Forms

$$a + bi = r(\cos\theta + i\sin\theta)$$

polar $\rightarrow$ rectangular   rectangular $\rightarrow$ polar

$$a = r\cos\theta$$ $$r = \sqrt{a^2 + b^2}$$

$$b = r\sin\theta$$ $$\tan\theta = \frac{b}{a}, \quad a \neq 0$$

Example 2. Converting Polar Form to Rectangular Form

Convert $z = 4\left(\cos\dfrac{\pi}{3} + i\sin\dfrac{\pi}{3}\right)$ to the rectangular form.

Solution.   Find the trigonometric values using the unit circle or other methods, then simplify the expression.

$$z = 4\left(\cos\frac{\pi}{3} + i\sin\frac{\pi}{3}\right)$$

$$= 4\left(\frac{1}{2} + i\frac{\sqrt{3}}{2}\right) \qquad\qquad \text{(Special trigonometric values)}$$

$$= 2 + 2\sqrt{3}i. \qquad\qquad\qquad \text{(Simplify.)}$$

Practice Now.   Convert the polar form to the rectangular form.

a) $4\left(\cos\dfrac{11\pi}{6} + i\sin\dfrac{11\pi}{6}\right)$ **b)** $\cos\pi + i\sin\pi$

Example 3. Converting Rectangular Form to Polar Form

Convert the complex number $3 - 2i$ to the polar form.

Solution. First identity the real part $a = 3$, and the imaginary part $b = -2$ (it is not $-2i$). Then

$$r = \sqrt{3^2 + (-2)^2} = \sqrt{13}, \quad \text{and} \quad \tan \theta = \frac{-2}{3}.$$

The complex number is in quadrant IV, and $\tan^{-1}(-2/3) = -33.7°$ is also in quadrant IV. Make the angle positive by adding $360°$ to get $\theta = -33.7° + 360° = 326.3°$. Then the answer is

$$z = \sqrt{13}\,(\cos 326.3° + i \sin 326.3°)$$

Practice Now. Convert the rectangular form to the polar form.

  a) $2 + i$ **b)** $-6 - 6i$

▶ Products, Quotients, and Powers in Polar Form

When we multiply two complex numbers, it is much easier if they are in polar form than in rectangular form.

✏ **Product and Quotient of Complex Numbers in Polar Form**

If $z = r(\cos A + i \sin A)$ and $w = s(\cos B + i \sin B)$, then

$$zw = rs\,[\cos(A + B) + i \sin(A + B)],$$
$$\frac{z}{w} = \frac{r}{s}\,[\cos(A - B) + i \sin(A - B)].$$

Simply put, we multiply or divide moduli, and we add or subtract arguments. These rules are derived from the sum and difference identities for sine and cosine.

Example 4. Multiplying or Dividing Complex Numbers in Polar Form

Let $z = 2(\cos 15° + i \sin 15°)$ and $w = 6(\cos 75° + i \sin 75°)$. Find zw and z/w. Answer in rectangular form.

Solution. For zw, multiply moduli: $2 \times 6 = 12$, and add angles: $15° + 75° = 90°$. Then

$$zw = 12(\cos 90° + i \sin 90°) = 12(0 + i \cdot 1) = 12i.$$

For z/w, divide moduli: $2/6 = 1/3$, and subtract angles: $15° - 75° = -60°$. Then

$$\frac{z}{w} = \frac{1}{3}[\cos(-60°) + i \sin(-60°)] = \frac{1}{3}\left[\frac{1}{2} + i\left(-\frac{\sqrt{3}}{2}\right)\right] = \frac{1}{6} - \frac{\sqrt{3}}{6}i.$$

Practice Now. Find zw and z/w. Answer in rectangular form.

  a) $z = 9(\cos 225° + i \sin 225°)$ **b)** $z = \cos 60° + i \sin 60°$
       $w = 3(\cos 45° + i \sin 45°)$ $w = 5(\cos 240° + i \sin 240°)$

DeMoivre's theorem in the following formula box is derived from the product and quotient formulas because powering is repeated multiplication (if $n > 0$) or repeated division (if $n < 0$).

DeMoivre's Theorem (Powers in Polar Form)

If $z = r(\cos\theta + i\sin\theta)$, then for any interger n,

$$z^n = r^n[\cos(n\theta) + i\sin(n\theta)].$$

Example 5. Taking Powers of a Complex Number in Polar Form

Let $z = 2\left(\cos\dfrac{\pi}{8} + i\sin\dfrac{\pi}{8}\right)$. Find z^4. Express the answer in rectangular form.

Solution.   Use DeMoivre's theorem to find the modulus: $2^4 = 16$, and the argument: $4 \times \dfrac{\pi}{8} = \dfrac{\pi}{2}$. Then

$$z^4 = 16\left(\cos\dfrac{\pi}{2} + i\sin\dfrac{\pi}{2}\right) = 16(0 + i) = 16i.$$

Practice Now.   Find each power. Express the answer in rectangular form.

a) $z = 4\left(\cos\dfrac{\pi}{4} + i\sin\dfrac{\pi}{4}\right);\quad z^3$
$\qquad\qquad\qquad$ **b)** $w = \cos 120° + i\sin 120°;\quad w^5$

Example 6. Taking Powers after Converting to the Polar Form

Let $z = -1 + i$. Use DeMoivre's theorem to find z^6.

Solution.   Convert the complex number into polar form so that we can apply DeMoivre's theorem. Identify the real part $a = -1$ and the imaginary part $b = 1$. Then the modulus r and argument θ of z satisfy

$$r = \sqrt{(-1)^2 + 1^2} = \sqrt{2}, \quad \tan\theta = \dfrac{1}{-1}.$$

Because z is in quadrant II and $\tan^{-1}(1/-1) = -45°$ is in quadrant IV, we adjust the angle by adding $180°$ and get $\theta = -45° + 180° = 135° = \dfrac{3\pi}{4}$. Now the modulus and argument of the power z^6 are

$$\text{modulus:}\quad r^6 = \left(\sqrt{2}\right)^6 = 8, \qquad\qquad \text{argument:}\quad 6\theta = 6\left(\dfrac{3\pi}{4}\right) = \dfrac{18\pi}{4} = \dfrac{9\pi}{2}.$$

Now apply DeMoivre's theorem.

$$z^6 = 8\left(\cos\dfrac{9\pi}{2} + i\sin\dfrac{9\pi}{2}\right) = 8\left(\cos\dfrac{\pi}{2} + i\sin\dfrac{\pi}{2}\right) = 8(1 + 0i) = 8$$

Practice Now.   Find each power using DeMoivre's theorem.

a) $z = -1 + \sqrt{3}i;\quad z^6$
$\qquad\qquad\qquad\qquad$ **b)** $w = 2 + 2i;\quad w^3$

▶ The nth Roots of Complex Numbers

For an integer $n > 0$, the nth roots of a complex number A are the solutions of the equation $z^n = A$. The number of solutions of such an equation is the same as the degree n, and the solutions are most easily described if A is in the polar form.

🖊 The nth Roots of a Complex Number

The equation $z^n = r(\cos\theta + i\sin\theta)$ has n complex solutions given by

$$z_k = \sqrt[n]{r}\left(\cos\frac{\theta + 360°k}{n} + i\sin\frac{\theta + 360°k}{n}\right)$$

for $k = 0, 1, 2, \ldots, n-1$.

Example 7. Finding the nth Roots of a Complex Number in Polar Form

Find the fourth roots of $81(\cos 100° + i\sin 100°)$.

Solution. We are solving the following equation of degree 4.

$$z^4 = 81(\cos 100° + i\sin 100°)$$

Apply the formula to get the modulus and arguments of the solutions, as follows.

$$\text{modulus: } \sqrt[4]{81} = 3, \qquad \text{arguments: } \frac{100° + 360°k}{4} = 25° + 90°k$$

Then substitute k with four different values, starting with 0, to get the arguments and use them to write the roots.

$$
\begin{array}{llll}
k = 0 & \Rightarrow & 25° + 90°(0) = 25°, & z_0 = 3(\cos 25° + i\sin 25°), \\
k = 1 & \Rightarrow & 25° + 90°(1) = 115°, & z_1 = 3(\cos 115° + i\sin 115°), \\
k = 2 & \Rightarrow & 25° + 90°(2) = 205°, & z_2 = 3(\cos 205° + i\sin 205°), \\
k = 3 & \Rightarrow & 25° + 90°(3) = 295°, & z_3 = 3(\cos 295° + i\sin 295°).
\end{array}
$$

Practice Now. Find the specified roots of each complex number.

a) cube roots of $64(\cos 120° + i\sin 120°)$

b) sixth roots of $9(\cos 90° + i\sin 90°)$

Example 8. Finding the nth Roots of a Complex Number

Find cube roots of $8i$. Answer in rectangular form.

Solution. First, convert the complex number into the polar form.

From the plot of the complex number on the right, obtain the modulus 8 and the argument $90°$.

$$8i = 8(\cos 90° + i \sin 90°)$$

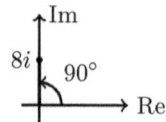

Apply the nth roots formula with $n = 3$ to find the modulus and arguments of the roots.

$$\text{modulus: } \sqrt[3]{8} = 2, \qquad \text{arguments: } \frac{90° + 360°k}{3} = 30° + 120°k$$

Then substitute k with three different values, starting with 0, to get the arguments. For $k = 0, 1$, and 2, the arguments are $30°, 150°$, and $270°$. Finally, list the roots.

$$z_0 = 2(\cos 30° + i \sin 30°) = 2\left(\frac{\sqrt{3}}{2} + \frac{1}{2}i\right) = \sqrt{3} + i,$$

$$z_1 = 2(\cos 150° + i \sin 150°) = 2\left(-\frac{\sqrt{3}}{2} + \frac{1}{2}i\right) = -\sqrt{3} + i,$$

$$z_2 = 2(\cos 270° + i \sin 270°) = 2\left(0 + (-1)i\right) = -2i.$$

Practice Now.   Find the specified roots of each complex number. Answer in rectangular form.

a) square roots of $9i$ **b)** fourth roots of -16

Exercises 4.5

(1–8) Plot the complex number in the complex plane and convert it to the rectangular form.

1. $5(\cos 180° + i \sin 180°)$ **2.** $6(\cos 150° + i \sin 150°)$

3. $\cos 225° + i \sin 225°$ **4.** $2(\cos 420° + i \sin 420°)$

5. $9\left(\cos \frac{3\pi}{4} + i \sin \frac{3\pi}{4}\right)$ **6.** $3\left(\cos 2\pi + i \sin 2\pi\right)$

7. $4\left(\cos \frac{5\pi}{3} + i \sin \frac{5\pi}{3}\right)$ **8.** $\cos \frac{7\pi}{6} + i \sin \frac{7\pi}{6}$

(9–16) Plot the complex number in the complex plane and convert it to the polar form.

9. $3 - 3i$ **10.** $\sqrt{3} + i$

11. $-2i$ **12.** 4

13. $-12 + 5i$ **14.** $4 - 3i$

15. $-2 - 4i$ **16.** $5 + 3i$

(17–22) Find zw and z/w.

17. $z = 6(\cos 50° + i \sin 50°)$, $w = 3(\cos 30° + i \sin 30°)$

18. $z = 2(\cos 200° + i \sin 200°)$, $w = 7(\cos 40° + i \sin 40°)$

19. $z = 6(\cos 70° + i \sin 70°)$, $w = 8(\cos 130° + i \sin 130°)$

20. $z = \cos 110° + i \sin 110°$, $w = 5(\cos 80° + i \sin 80°)$

21. $z = 4\left(\cos \frac{7\pi}{3} + i \sin \frac{7\pi}{3}\right)$

     $w = 2\left(\cos \frac{2\pi}{3} + i \sin \frac{2\pi}{3}\right)$

22. $z = 5\left(\cos \frac{5\pi}{4} + i \sin \frac{5\pi}{4}\right)$

     $w = 3\left(\cos \pi + i \sin \pi\right)$

(23–26) Find the power. Express the answer in the polar form.

23. $[5(\cos 40° + i \sin 40°)]^2$ **24.** $[2(\cos 100° + i \sin 100°)]^3$

25. $\left(\cos \frac{\pi}{5} + i \sin \frac{\pi}{5}\right)^2$ **26.** $\left(\cos \frac{3\pi}{10} + i \sin \frac{3\pi}{10}\right)^4$

(27–36) Find the power. Express the answer in the rectangular form.

27. $[2(\cos 50° + i \sin 50°)]^6$ **28.** $[3(\cos 135° + i \sin 135°)]^2$

29. $(\cos 150° + i \sin 150°)^3$ **30.** $\left(\cos \frac{\pi}{3} + i \sin \frac{\pi}{3}\right)^5$

31. $\left[4\left(\cos \frac{\pi}{4} + i \sin \frac{\pi}{4}\right)\right]^4$ **32.** $\left[2\left(\cos \frac{5\pi}{6} + i \sin \frac{5\pi}{6}\right)\right]^4$

33. $(2 + 2i)^3$ **34.** $\left(\frac{3\sqrt{3}}{2} - \frac{3}{2}i\right)^4$

35. $\left(-\dfrac{1}{2} + \dfrac{\sqrt{3}}{2}i\right)^6$ **36.** $(2i)^8$

(37–40) Find the nth roots. Express the answer in the polar form.

37. cube roots of $27(\cos 150° + i\sin 150°)$

38. fourth roots of $64(\cos 320° + i\sin 320°)$

39. square roots of $3\left(\cos\dfrac{\pi}{4} + i\sin\dfrac{\pi}{4}\right)$

40. sixth roots of $\cos\dfrac{3\pi}{2} + i\sin\dfrac{3\pi}{2}$

(41–46) Find the nth roots. Express the answer in the rectangular form.

41. square roots of $49(\cos 60° + i\sin 60°)$

42. cube roots of $8(\cos 90° + i\sin 90°)$

43. square roots of $-1 + \sqrt{3}i$

44. fourth roots of -4

45. cube roots of $-64i$

46. sixth roots of 1

Chapter **5**

SEQUENCES AND SERIES

5.1   Notations for Sequences and Series

▶ **Sequences and Patterns**

A **sequence** is an ordered list of numbers. For example,

$$1, 3, 5, 7, \ldots \qquad \text{is the sequence of odd numbers,}$$
$$1, 4, 9, 16, \ldots \qquad \text{is the sequence of squared numbers.}$$

Each number in a sequence is called a **term**. For example, in the sequence of odd numbers, 1 is the first term, 3 is the second term, and so on. The nth term is denoted by a_n where a is the name of the sequence and n is the index representing the position of the term in the sequence. The nth term a_n is also called the **general term** of the sequence.

Example 1. Finding Specific Terms from the General Term

Find the 9th and 10th terms of each sequence.

a) $a_n = 2n + 1$

b) $b_n = n^2$

Solution.   To find specific terms, we replace n with the given number.

a) 9th term: $a_9 = 2(9) + 1 = 19$
   10th term: $a_{10} = 2(10) + 1 = 21$

b) 9th term: $b_9 = 9^2 = 81$
   10th term: $b_{10} = 10^2 = 100$

Practice Now.   Find the first four terms of the sequence.

a) $a_n = n^2 - 4$

b) $b_n = \dfrac{1}{2n + 1}$

Typically, we are interested in sequences that have patterns. If a sequence has a pattern, we can find the expression for the general term a_n by analyzing the pattern carefully.

Example 2. Finding the Generel Term from the Pattern

Find the general term a_n of the sequence $\dfrac{2}{1}, \dfrac{3}{2}, \dfrac{4}{4}, \dfrac{5}{8}, \dfrac{6}{16}, \ldots$

Solution.   The numerators progress linearly, and the denominators progress exponentially. We examine the relationship between these numbers and indices as follows. First, rewrite the denominators in power form together with the a_n notation.

$$a_1 = \frac{2}{2^0}, \quad a_2 = \frac{3}{2^1}, \quad a_3 = \frac{4}{2^2}, \quad a_4 = \frac{5}{2^3}, \quad a_5 = \frac{6}{2^4}, \quad \ldots$$

Then analyze the numbers. Observe that the numerators are (index)$+1$,

$$a_1 = \frac{2}{2^0}, \quad a_2 = \frac{3}{2^1}, \quad a_3 = \frac{4}{2^2}, \quad a_4 = \frac{5}{2^3}, \quad a_5 = \frac{6}{2^4}, \quad \ldots$$

and that the exponents in the denominators are (index)-1.

$$a_1 = \frac{2}{2^0}, \quad a_2 = \frac{3}{2^1}, \quad a_3 = \frac{4}{2^2}, \quad a_4 = \frac{5}{2^3}, \quad a_5 = \frac{6}{2^4}, \quad \ldots$$

Then write the answer.

$$a_n = \frac{n+1}{2^{n-1}}.$$

Practice Now.   Find the general term of the following sequences.

a) $\sqrt{3}, \sqrt{4}, \sqrt{5}, \sqrt{6}, \ldots$ **b)** $3, 9, 27, 81, 243, \ldots$

▶ Recursively Defined Sequences

A **recursively defined sequence** is a sequence defined by one or more initial terms and recursive formulas that describe each term of the sequence in terms of preceding terms. For example, in the recursive formula $a_{n+1} = a_n + 3$, the terms a_{n+1} and a_n represent the next term and the current term, respectively. Using the formula, we can compute the next term by adding 3 to the current term.

It feels like playing dominos when one computes terms of a recursively defined sequence. The first term leads to the second term, the second term leads to the third term, and so forth.

Example 3. Finding Terms of a Recursively Defined Sequence

Find the first five terms of the sequence defined by

$$a_1 = 4, \quad a_{n+1} = 2a_n - 3 \quad \text{for } n \geq 1.$$

Solution.   The first term is given. Substitute n in the recursive formula with a specific value and get a formula to compute a specific term.

$$a_1 = 4$$

$$n = 1: \quad a_{1+1} = 2a_1 - 3 \quad \Rightarrow \quad a_2 = 2a_1 - 3 = 2(4) - 3 = 5,$$

$$n = 2: \quad a_{2+1} = 2a_2 - 3 \quad \Rightarrow \quad a_3 = 2a_2 - 3 = 2(5) - 3 = 8,$$

$$n = 3: \quad a_{3+1} = 2a_3 - 3 \quad \Rightarrow \quad a_4 = 2a_3 - 3 = 2(8) - 3 = 13,$$

$$n = 4: \quad a_{4+1} = 2a_4 - 3 \quad \Rightarrow \quad a_5 = 2a_4 - 3 = 2(13) - 3 = 23.$$

Practice Now. Find the first four terms of the recursively defined sequence.

$$a_1 = 3, \quad a_{n+1} = 3a_n - 7 \quad \text{for } n \geq 1.$$

▶ **Factorial Notation**

> ✏️ **Factorials**
>
> For any positive integer n, the n **factorial** is written as $n!$ and is defined by
> $$n! = n(n-1)(n-2)\cdots 3 \cdot 2 \cdot 1.$$
> If $n = 0$, then zero factorial is defined by $0! = 1$. (Warning: zero factorial is not zero.)

To compute the factorial, we begin with the given number n and decrease it one by one until we reach 1. Then we multiply them all. For example, $4! = 4 \cdot 3 \cdot 2 \cdot 1 = 24$. It works in the same way with variables as well. For example,

$$(n+2)! = (n+2)(n+1)n(n-1)\cdots 3 \cdot 2 \cdot 1,$$
$$(n-1)! = (n-1)(n-2)(n-3)\cdots 3 \cdot 2 \cdot 1,$$
$$p! = p(p-1)(p-2)(p-3)\cdots 3 \cdot 2 \cdot 1.$$

The recursive formula for $n!$ is

$$a_1 = 1, \quad a_{n+1} = (n+1)a_n \quad \text{for } n \geq 1.$$

Example 4. Simplifying Numeric Factorials

Simplify $\dfrac{7!}{9!}$.

Solution. Apply the definition, then cancel common factors to find the answer.

$$\frac{7!}{9!} = \frac{7 \cdot 6 \cdot 5 \cdot 4 \cdot 3 \cdot 2 \cdot 1}{9 \cdot 8 \cdot 7 \cdot 6 \cdot 5 \cdot 4 \cdot 3 \cdot 2 \cdot 1}$$
$$= \frac{\cancel{7} \cdot \cancel{6} \cdot \cancel{5} \cdot \cancel{4} \cdot \cancel{3} \cdot \cancel{2} \cdot \cancel{1}}{9 \cdot 8 \cdot \cancel{7} \cdot \cancel{6} \cdot \cancel{5} \cdot \cancel{4} \cdot \cancel{3} \cdot \cancel{2} \cdot \cancel{1}}$$

$$= \frac{1}{9 \cdot 8}$$
$$= \frac{1}{72}$$

Practice Now.   Simplify each expression.

a) $\dfrac{4!}{6!}$

b) $\dfrac{7!}{2!\,5!}$

Example 5. Simplifying Symbolic Factorials

Simplify $\dfrac{(n-1)!}{(n+1)!}$.

Solution.   Apply the definition and cancel common factors, then simplify the expression.

$$\frac{(n-1)!}{(n+1)!} = \frac{(n-1)(n-2)\cdots 3 \cdot 2 \cdot 1}{(n+1)n(n-1)(n-2)\cdots 3 \cdot 2 \cdot 1}$$

$$= \frac{\cancel{(n-1)}\cancel{(n-2)}\cdots\cancel{3}\cdot\cancel{2}\cdot\cancel{1}}{(n+1)n\cancel{(n-1)}\cancel{(n-2)}\cdots\cancel{3}\cdot\cancel{2}\cdot\cancel{1}}$$

$$= \frac{1}{(n+1)n}$$

$$= \frac{1}{n^2 + n}$$

Practice Now.   Simplify $\dfrac{(n+2)!}{n!}$.

▶ Series and Summation Notation

A **series** is the sum of terms of a sequence. The notation $\displaystyle\sum_{i=p}^{q} a_i$ denotes the sum from the pth term all the way up to the qth term. In mathematical formula,

$$\sum_{i=p}^{q} a_i = a_p + a_{p+1} + \cdots + a_{q-1} + a_q.$$

In this notation, i is the index of the sequence, p is the lower limit, and q is the upper limit of the index. We may use other letters, such as n, k, or j, instead of i for the index.

Example 6. Evaluating Summation Notation

Evaluate the sum $\displaystyle\sum_{k=2}^{5}(k^2 - 1)$.

Solution.   Take the sum of the numbers obtained by substituting k with the numbers from the lower limit 2 to the upper limit 5, increasing the index one by one.

$$\sum_{k=2}^{5}(k^2 - 1) = (2^2 - 1) + (3^2 - 1) + (4^2 - 1) + (5^2 - 1)$$

$$= 3 + 8 + 15 + 24$$
$$= 50$$

Practice Now.   Evaluate the sum $\displaystyle\sum_{j=0}^{5}(3j + 1)$.

Example 7. Expressing the Sum in Summation Notation

Express the following sum in summation notation.

$$\frac{1}{2} + \frac{2}{3} + \frac{3}{4} + \frac{4}{5} + \cdots + \frac{24}{25}$$

Solution.   We begin with finding the general term expressing individual terms.

$$a_1 = \frac{1}{2}, \quad a_2 = \frac{2}{3}, \quad a_3 = \frac{3}{4}, \quad a_4 = \frac{4}{5}, \quad \ldots, \quad a_? = \frac{24}{25}$$

Observe the pattern that the numerators are eqaul to the indices and that the denominators are bigger than the numerators by exactly 1. Therefore, $a_n = \dfrac{n}{n+1}$. The last term $\dfrac{24}{25}$ is for $n = 24$. It will be the last index. Therfore, the answer is

$$\sum_{n=1}^{24} \frac{n}{n+1}.$$

Practice Now.   Express the sum in summation notation.

a) $2 + 2^2 + 2^3 + \cdots + 2^{10}$

b) $\dfrac{2}{2} + \dfrac{2}{3} + \dfrac{2}{4} + \cdots + \dfrac{2}{100}$

Exercises 5.1

(1–8) Find the first four terms of the sequence.

1. $a_n = 5n + 3$

2. $b_n = (-1)^n$

3. $c_n = n^2 - 5n + 4$

4. $d_n = \dfrac{n}{n+2}$

5. $e_n = 5$

6. $f_n = n!$

7. $g_n = \dfrac{(-1)^{n+1}}{n^2}$

8. $h_n = 3^n - 1$

(9–18) Find the general term a_n of the sequence.

9. $1, 2, 3, 4, 5, \ldots$

10. $2, 3, 4, 5, 6, \ldots$

11. $\dfrac{2}{3}, \dfrac{3}{4}, \dfrac{4}{5}, \dfrac{5}{6}, \ldots$

12. $1 + \dfrac{1}{2}, 1 + \dfrac{1}{3}, 1 + \dfrac{1}{4}, 1 + \dfrac{1}{5}, \ldots$

13. $2, 4, 6, 8, 10, \ldots$

14. $0, 3, 6, 9, 12, \ldots$

15. $-1, 1, -1, 1, -1, 1, \ldots$

16. $1, -1, 1, -1, 1, -1, \ldots$

17. $2, 4, 8, 16, \ldots$

18. $\dfrac{-1}{2}, \dfrac{1}{4}, \dfrac{-1}{8}, \dfrac{1}{16}, \ldots$

(19–23) Find the first four terms of the recursively defined sequence.

19. $a_1 = 1, a_{n+1} = a_n + 3$

20. $a_1 = 2, a_{n+1} = 2a_n$

21. $a_1 = 7, a_{n+1} = -2a_n + 1$

22. $a_1 = 3, a_{n+1} = \dfrac{1}{a_n}$

23. $a_1 = 1, a_{n+1} = na_n$

(24–31) Simplify the factorial notation.

24. $6!$

25. $\dfrac{10!}{7!}$

26. $\dfrac{3!}{5!}$

27. $\dfrac{6!}{4!\,2!}$

28. $\dfrac{8!}{5!\,3!}$

29. $\dfrac{(n+1)!}{n!}$

30. $\dfrac{(n+1)!}{(n+2)!}$

31. $\dfrac{(n-2)!}{(n-4)!}$

(32–36) Find the sum.

32. $\displaystyle\sum_{k=1}^{4} k^2$

33. $\displaystyle\sum_{j=0}^{5} (2j-1)$

34. $\displaystyle\sum_{i=1}^{4} 5$

35. $\displaystyle\sum_{n=1}^{4} \dfrac{2}{n}$

36. $\displaystyle\sum_{k=1}^{4} (-1)^{k+1} k$

(37–41) Express the sum in summation notation.

37. $3 + 4 + 5 + 6 + \cdots + 100$

38. $\dfrac{1}{2} + \dfrac{2}{3} + \dfrac{3}{4} + \dfrac{4}{5} + \cdots + \dfrac{10}{11}$

39. $2 + 4 + 6 + 8 + \cdots + 54$

40. $\dfrac{1}{2} + \dfrac{1}{2\cdot 3} + \dfrac{1}{3\cdot 4} + \dfrac{1}{4\cdot 5} + \cdots + \dfrac{1}{9\cdot 10}$

41. $1 - 2 + 4 - 8 + \cdots + 256$

5.2　Arithmetic Sequences and Their Sums

▶ Arithmetic Sequences

An **arithmetic sequence** is a sequence of numbers where each term after the first is found by adding the previous term by a constant d called the **common difference**. In recursive formula,

$$a_{n+1} = a_n + d \qquad \text{for } n \geq 1.$$

For example, the following sequence is an arithmetic sequence with the common difference $d = 3$.

$$\overset{+3}{\frown}\ \overset{+3}{\frown}\ \overset{+3}{\frown}\ \overset{+3}{\frown}\ \overset{+3}{\frown}$$
$$11,\quad 14,\quad 17,\quad 20,\quad 23,\quad 26,\quad \ldots$$

✏️ **The General Term of an Arithmetic Sequence**

The general term of an arithmetic sequence with the first term a_1 and common difference d is

$$a_n = a_1 + (n-1)d \qquad \text{for } n \geq 1.$$

We may substitute n with an index value to find a specific term, provided the first term a_1 and the common difference d are given. For example, if $a_1 = 11$ and $d = 3$, then the 10th term a_{10} is found by

$$a_{10} = a_1 + (10-1)d = a_1 + 9d = 11 + 9(3) = 38.$$

Example 1. Finding the General Term of an Arithmetic Sequence

Find the general term a_n of the arithmetic sequence. Then find the 20th term of the sequence.

$$3, 5, 7, 9, \ldots$$

Solution. We notice that $a_1 = 3$ and $d = 2$. Therefore,

$$a_n = 3 + (n-1)2 = 3 + 2n - 2 = 2n + 1.$$

Then the 20th term is $a_{20} = 2(20) + 1 = 41$.

Practice Now. Find the general term of the given arithmetic sequence, then find the 10th term.

a) $4, 9, 14, 19, \ldots$ **b)** $50, 48, 46, 44, \ldots$

Example 2. Finding the Number of Terms

How many terms are there in the arithmetic sequence 2, 5, 8, 11, $\ldots$, 77?

Solution. Note that the first term is $a_1 = 2$ and the common difference is $d = 3$. We assume the last term is the nth term where n is unknown; that is, $a_n = 77$. Using the general term formula, we have

$$a_n = a_1 + (n-1)d = 2 + (n-1)3 = 2 + 3n - 3 = 3n - 1$$

Now substitute the value of a_n and solve the equation $77 = 3n - 1$ to get $n = 26$. We conclude that the last term is the 26th term; that is, there are a total of 26 terms in the given sequence.

Practice Now. How many terms are there in each arithmetic sequence?

a) $5, 7, 9, \ldots, 107$ **b)** $32, 29, 26, \ldots, -10$

Given two specific terms of an arithmetic sequence, we can find the common difference using a formula, which is reminiscent of the slope of a line formula. The reason is that d is the rate at which the terms increase as the index changes.

> ### ✎ The Common Difference of an Arithmetic Sequence
>
> The common difference d of an arithmetic sequence with two terms a_m and a_n is
> $$d = \frac{a_m - a_n}{m - n}.$$

Example 3. Finding the Common Difference

Find the common difference d of the arithmetic sequence a_n given $a_4 = 16$ and $a_{17} = 42$.

Solution. If we use the above formula, the answer is

$$d = \frac{a_{17} - a_4}{17 - 4} = \frac{42 - 16}{17 - 4} = \frac{26}{13} = 2.$$

Practice Now. Find the common difference of each arithmetic sequence.

a) $a_4 = 23, \quad a_{10} = 5$ **b)** $b_1 = 5, \quad b_{10} = 23$

▶ Arithmetic Series—The Sum of an Arithmetic Sequence

The sum of a finite arithmetic sequence can be computed easily with the first term, the last term, and the number of terms.

📝 **Arithmetic Sum**

The sum of the first n terms of an arithmetic sequence $a_1, a_2, \ldots$ is

$$S_n = \frac{n(a_1 + a_n)}{2} = \frac{n[a_1 + a_1 + (n-1)d]}{2}$$

where n is the number of terms, a_1 is the first term, and a_n is the last term.

Example 4. Evaluating Arithmetic Sum Given the First and Last Terms

Find the sum $3 + 7 + 11 + 15 + \cdots + 55$.

Solution.   To use the above arithmetic series formula, we need n, a_1, and a_n. The first term $a_1 = 3$ and the last term $a_n = 55$ are given, but the number of terms n is unkown. Find it using the general term formula

$$a_n = a_1 + (n-1)d.$$

From the pattern we have $d = 4$. Now substitute known values.

$$55 = 3 + (n-1)(4)$$

Then solve the equation to get $n = 14$. Finally, apply the summation formula.

$$S_{14} = \frac{14(3 + 55)}{2} = 406$$

Practice Now.   Find the sum of arithmetic sequence.

a) $5 + 6 + 7 + \cdots + 26$ b) $7 + 9 + 11 + \cdots + 79$

Example 5. Evaluating Arithmetic Sum Given the Number of Terms

Find the sum of the first 30 terms of the arithmetic sequence $40, 36, 32, \ldots$.

Solution.   To use the arithmetic series formula, we need n, a_1, and a_n. The first term $a_1 = 40$ and the number of terms $n = 30$ are given, but the last term a_{30} is unknown. Find it using the general term formula

$$a_n = a_1 + (n-1)d.$$

From the pattern we see that $d = -4$. Now substitute known values.

$$a_{30} = 40 + (30 - 1)(-4) = -76.$$

Finally, apply the summation formula with $n = 30$ to find the sum.

$$S_{30} = \frac{30(a_1 + a_{30})}{2} = \frac{30(40 + (-76))}{2} = -540$$

Practice Now.   Find the sum of the first 40 terms of the arithmetic sequence.

a) $2, 6, 10, 14, \ldots$ **b)** $3, 6, 9, 12, \ldots$

Exercises 5.2

(1–7) Determine whether the sequence is arithmetic or not. If it is arithmetic, find the first term a_1 and common difference d.

1. $1, 3, 5, 7, 9, \ldots$

2. $5, 2, -1, -4, -7, \ldots$

3. $2, 4, 8, 16, 32, \ldots$

4. $2, 3, 5, 8, 12, \ldots$

5. $1.2, 2.2, 3.2, 4.2, \ldots$

6. $\frac{1}{2}, 1, \frac{3}{2}, 2, \frac{5}{2} \ldots$

7. $\pi, 2\pi, 3\pi, 4\pi, \ldots$

(8–13) Find the general term a_n of the arithmetic sequence. Then find the 10th term.

8. $2, 5, 8, 11, 14, \ldots$

9. $12, 6, 0, -6, -12, \ldots$

10. $1, 6, 11, 16, 21, \ldots$

11. $\frac{1}{2}, \frac{3}{2}, \frac{5}{2}, \frac{7}{2}, \frac{9}{2}, \ldots$

12. $\frac{1}{4}, 1, \frac{7}{4}, \frac{5}{2}, \frac{13}{4}, \ldots$

13. $\frac{\pi}{2}, \pi, \frac{3\pi}{2}, 2\pi, \frac{5\pi}{2}, \ldots$

(14–18) Find the common difference d and the expression for a_n.

14. $a_3 = 12$, $a_{10} = 40$

15. $a_4 = 7$, $a_{12} = -17$

16. $a_3 = 5$, $a_{20} = 39$

17. $a_7 = 10$, $a_{13} = 40$

18. $a_3 = 2$, $a_{13} = 7$

(19–22) How many terms are there in the arithmetic sequence?

19. $1, 3, 5, \ldots, 59$

20. $4, 7, 10, \ldots, 61$

21. $2, -1, -4, \ldots, -22$

22. $\frac{1}{2}, \frac{3}{4}, 1, \ldots, 3$

(23–26) Find the sum of the arithmetic sequence.

23. $\displaystyle\sum_{n=1}^{10}(3n + 1)$

24. $\displaystyle\sum_{k=1}^{100}(k + 2)$

25. $\displaystyle\sum_{n=1}^{200}(-n)$

26. $\displaystyle\sum_{i=5}^{13}(30 - 2i)$

(27–29) Find the sum of the first n terms of the arithmetic sequence.

27. $1, 6, 11, 16, \ldots; n = 30$

28. $-15, -12, -9, \ldots; n = 15$

29. $5, 8, 11, 14, \ldots; n = 20$

(30–35) Find the sum of the arithmetic sequence.

30. $1 + 2 + 3 + \cdots + 30$

31. $2 + 4 + 6 + \cdots + 100$

32. $3 + 7 + 11 + \cdots + 99$

33. $10 + 9 + 8 + \cdots + (-9)$

34. $7 + 12 + 17 + \cdots + 107$

35. $\frac{1}{3} + 1 + \frac{5}{3} + \cdots + 5$

5.3   Geometric Sequences and Series

▶ Geometric Sequences

A **geometric sequence** is a sequence of numbers where each term after the first is found by multiplying the previous term by a nonzero constant r called the **common ratio**.

For example, the following sequence is a geometric sequence with the common ratio $r = 3$.

$$\begin{array}{ccccccc} \times 3 & \times 3 & \times 3 & \times 3 & \times 3 & & \\ 1, & 3, & 9, & 27, & 81, & 243, & \ldots \end{array}$$

We call r the common ratio because it is the ratio of any two consecutive terms. For example, in the above sequence, $9/3 = 3$, $81/27 = 3$, and so on.

The recursive formula of a geometric sequence is

$$a_{n+1} = ra_n \qquad \text{for } n \geq 1.$$

> ### ✎ The General Term of a Geometric Sequence
>
> The general term of the geometric sequence with the first term a and common ratio r is
> $$a_n = ar^{n-1} \qquad \text{for } n \geq 1.$$

Example 1. Finding the General Term of a Geometric Sequence

Find the general term a_n of the geometric sequence $\dfrac{2}{3}, 2, 6, 18, 54, \ldots$.

Solution. Note that the first term is $a = \dfrac{2}{3}$ and the common ratio is $r = 3$. Apply the general term formula to get

$$a_n = ar^{n-1} = \frac{2}{3}(3^{n-1}) = 2(3^{n-2}).$$

Practice Now. Find the general term of each geometric sequence.

a) $1, 2, 4, 8, \ldots$

b) $6, 2, \dfrac{2}{3}, \dfrac{2}{9} \ldots$

Example 2. Finding a Specific Term of a Geometric Sequence

Find the 12th term of the geometric sequence with the first term 10 and the common ratio 1.5. Round the answer to two decimal places.

Solution. The general term is $a_n = 10(1.5)^{n-1}$. Substitute $n = 12$ to find the 12th term.

$$a_{12} = 10(1.5)^{12-1} = 10(1.5)^{11} \approx 864.98$$

Practice Now. Find the 10th term of the geometric sequence with the first term a_1 and common ratio r. Round the answer to two decimal places.

a) $a_1 = 500$, $r = 0.98$

b) $a_1 = 40$, $r = 1.2$

► Finite Geometric Series

A **geometric series** is the sum of a geometric sequence. A finite geometric series is the sum of a finite number of geometric terms.

$$\sum_{k=1}^{n} ar^{k-1} = a + ar + ar^2 + \cdots + ar^{n-1}$$

> ### Finite Geometric Series
>
> The sum S_n of the first n terms of the geometric sequence with the first term a and common ratio r is
>
> $$S_n = \frac{a(1 - r^n)}{1 - r} = \frac{a(r^n - 1)}{r - 1}.$$

The first formula is convenient when $r < 1$, and the second formula is so when $r > 1$, simply because denominators will be positive.

Example 3. Finding Finite Geometric Series in Summation Notation

Find the sum $\sum_{n=1}^{10} 70,000(1.05)^{n-1}$.

Solution. Compare the general term formula ar^{n-1} with the problem, and note that $a = 70,000$ and $r = 1.05$. Also note that the number of terms is $n = 10$. Then apply the finite geometric series formula.

$$S_{10} = \frac{70,000(1.05^{10} - 1)}{1.05 - 1}$$
$$\approx 88,0452.48.$$

Practice Now. Find the sum of each geometric series. Round to the nearest hundredth.

a) $\sum_{n=1}^{5} 3000(1.1)^{n-1}$

b) $\sum_{n=1}^{20} 400(0.98)^{n-1}$

► Infinite Geometric Series

An **infinite geometric series** is the sum of infinitely many geometric terms. The infinite sum

$$a + ar + ar^2 + ar^3 + \cdots$$

either exists or does not exist. If $|r| < 1$, then the terms ar^{n-1} approach 0; that is, the terms keep getting closer to 0 as n increases. It happens that their sum exists and equals $\frac{a}{1 - r}$. On the other hand, if $|r| \geq 1$, then the terms ar^{n-1} increase (or stay the same) in magnitude as n increases. And the sum does not exist in this case.

> ### ✏ Infinite Geometric Series
>
> The infinite sum of the geometric sequence with the first term a and the common ratio r is
>
> $$\sum_{n=1}^{\infty} ar^{n-1} = \begin{cases} \dfrac{a}{1-r} & \text{if } |r| < 1, \\ \text{does not exist} & \text{if } |r| \geq 1. \end{cases}$$

Example 4. Finding Infinite Geometric Series

Find the sum of each geometric series.

a) $9 - 6 + 4 - \dfrac{8}{3} + \cdots$ **b)** $\dfrac{1}{2} + \dfrac{3}{4} + \dfrac{9}{8} + \cdots$ **c)** $\displaystyle\sum_{n=0}^{\infty} 2\left(\dfrac{1}{3}\right)^{n+1}$

Solution.

a) Note that the first term is $a = 9$, and the common ratio is $r = \dfrac{a_2}{a_1} = \dfrac{-6}{9} = -\dfrac{2}{3}$. Because $\left|-\dfrac{2}{3}\right| = \dfrac{2}{3} < 1$, the sum exists, and it is

$$\frac{a}{1-r} = \frac{9}{1 - \left(-\dfrac{2}{3}\right)} = \frac{9}{1 + \dfrac{2}{3}} = \frac{9}{\dfrac{5}{3}} = 9 \cdot \frac{3}{5} = \frac{27}{5}.$$

b) The common ratio is

$$r = \frac{a_2}{a_1} = \frac{3/4}{1/2} = \frac{3}{4} \cdot \frac{2}{1} = \frac{3}{2}.$$

Because $\left|\dfrac{3}{2}\right| > 1$, the sum does not exist.

c) The index starts with $n = 0$ rather than 1. In order to find the first term a, we compute the value of the expression at $n = 0$.

$$a = 2\left(\frac{1}{3}\right)^{0+1} = 2 \cdot \frac{1}{3} = \frac{2}{3}.$$

Next, the common ratio is the base $\dfrac{1}{3}$, which is less than 1. Therefore, the sum exists and its value is

$$\frac{a}{1-r} = \frac{\dfrac{2}{3}}{1 - \dfrac{1}{3}} = \frac{\dfrac{2}{3}}{\dfrac{2}{3}} = 1.$$

Practice Now.   Find the sum of each geometric sequence.

a) $10 + 2 + \dfrac{2}{5} + \dfrac{2}{25} + \cdots$ **b)** $2 - 4 + 8 - 16 + \cdots$ **c)** $\displaystyle\sum_{n=0}^{\infty} \left(\dfrac{2}{5}\right)^{n}$

Exercises 5.3

(1–7) Determine whether the sequence is geometric or not. If it is geometric, find the first term a_1 and common ratio r.

1. $2, 4, 8, 16, \ldots$

2. $2, 4, 6, 8, \ldots$

3. $18, 6, 2, \dfrac{2}{3}, \ldots$

4. $-3, 9, -27, 81, \ldots$

5. $1, -1, 1, -1, \ldots$

6. $2, -2, -2, 2, -2, -2, 2, \ldots$

7. $1, 3, 6, 10, 15, \ldots$

(8–17) Find the general term of the geometric sequence.

8. $1, 3, 9, 27, \ldots$

9. $2, 4, 8, 16, \ldots$

10. $1, -1, 1, -1, \ldots$

11. $-1, 1, -1, 1, \ldots$

12. $-3, -6, -12, -24, \ldots$

13. $1, -2, 4, -8, 16, \ldots$

14. $5, \dfrac{5}{3}, \dfrac{5}{9}, \dfrac{5}{27}, \ldots$

15. $3, \dfrac{6}{5}, \dfrac{12}{25}, \dfrac{24}{125}, \ldots$

16. $7, 7(1.1), 7(1.1)^2, 7(1.1)^3, \ldots$

17. $\pi, \pi^2, \pi^3, \pi^4, \ldots$

(18–25) Find the indicated term of the geometric sequence with the given information.

18. a_3 when $a_1 = 7$ and $r = 3$

19. a_7 when $a_1 = 6$ and $r = 2$

20. a_5 when $a_1 = 3$ and $r = -3$

21. a_5 when $a_1 = \dfrac{2}{49}$ and $r = 7$

22. a_4 when $a_1 = 1$ and $r = \dfrac{2}{3}$

23. a_5 when $a_1 = \dfrac{49}{5}$ and $r = \dfrac{5}{7}$

24. a_3 when $a_1 = 10$ and $r = 1.3$

25. a_3 when $a_1 = 50$ and $r = 0.5$

(26–31) Find the sum of the first n terms of the geometric sequence. Round to the nearest hundredth if necessary.

26. $3, 6, 12, 24, \ldots;\quad n = 6$

27. $1, 3, 9, 27, \ldots;\quad n = 6$

28. $\dfrac{3}{2}, \dfrac{9}{2}, \dfrac{27}{2}, \dfrac{81}{2}, \ldots;\quad n = 6$

29. $2, 2(0.8), 2(0.8)^2, 2(0.8)^3, \ldots;\quad n = 10$

30. $4, 4(1.1), 4(1.1)^2, 4(1.1)^3, \ldots;\quad n = 20$

31. $9, 9(0.95), 9(0.95)^2, 9(0.95)^3, \ldots;\quad n = 9$

(32–37) Find the finite geometric series. Round to the nearest hundredth if necessary.

32. $\displaystyle\sum_{n=1}^{5} 2(3)^{n-1}$

33. $\displaystyle\sum_{n=1}^{5} 4(2)^{n-1}$

34. $\displaystyle\sum_{n=1}^{6} 10(-2)^{n-1}$

35. $\displaystyle\sum_{n=1}^{9} 7(1.2)^{n-1}$

36. $\displaystyle\sum_{n=1}^{10} 20(0.7)^{n-1}$

37. $\displaystyle\sum_{n=1}^{7} 2500(-1.05)^{n-1}$

(38–52) Find the infinite geometric series.

38. $6 + 2 + \dfrac{2}{3} + \dfrac{2}{9} + \cdots$

39. $6 - 2 + \dfrac{2}{3} - \dfrac{2}{9} + \cdots$

40. $1 + 4 + 16 + 64 + \cdots$

41. $3 + \dfrac{6}{5} + \dfrac{12}{25} + \dfrac{24}{125} + \cdots$

42. $2 - \dfrac{3}{2} + \dfrac{9}{8} - \dfrac{27}{32} + \cdots$

43. $1.01 + (1.01)^2 + (1.01)^3 + (1.01)^4 + \cdots$

44. $4(0.9) + 4(0.9)^2 + 4(0.9)^3 + 4(0.9)^4 + \cdots$

45. $\displaystyle\sum_{n=1}^{\infty} 3\left(\dfrac{1}{2}\right)^{n-1}$

46. $\displaystyle\sum_{n=1}^{\infty} 5\left(\dfrac{3}{4}\right)^{n-1}$

47. $\displaystyle\sum_{n=0}^{\infty} \left(-\dfrac{1}{2}\right)^{n}$

48. $\displaystyle\sum_{n=1}^{\infty} \dfrac{5}{9}\left(\dfrac{3}{2}\right)^{n-1}$

49. $\displaystyle\sum_{n=1}^{\infty} 100(0.9)^{n-1}$

50. $\displaystyle\sum_{n=1}^{\infty} 100(-0.9)^{n-1}$

51. $\displaystyle\sum_{n=1}^{\infty} 9(1.02)^{n-1}$

52. $\displaystyle\sum_{n=1}^{\infty} 36(-0.8)^{n-1}$

Appendix A

Answers to Exercises

Chapter 1

▶ Section 1.1

1. $(-\infty, -2) \cup (-2, \infty)$

3. $(-\infty, 2) \cup (2, \infty)$

5. $(-\infty, 5) \cup (5, \infty)$

7. x-intercept:$(1, 0)$; y-intercept: $\left(0, -\dfrac{1}{2}\right)$

9. x-intercept:$(-2, 0)$, $(1, 0)$; y-intercept: $\left(0, -\dfrac{2}{7}\right)$

11. x-intercept:$\left(-\dfrac{4}{5}, 0\right)$; y-intercept: $\left(0, \dfrac{4}{25}\right)$

13. vertical asymptote: $x = -3$;
    horizontal asymptote: $y = 1$

15. vertical asymptote: $x = -1$ and $x = 6$;
    horizontal asymptote: $y = 1$

17. vertical asymptote: $x = 2$;
    horizontal asymptote: $y = 2$

19. vertical asymptote: $x = -5$ and $x = -3$;
    horizontal asymptote: x−axis $(y = 0)$

21. $y = x + 1$

23. no oblique asymptote

25. no oblique asymptote

27.

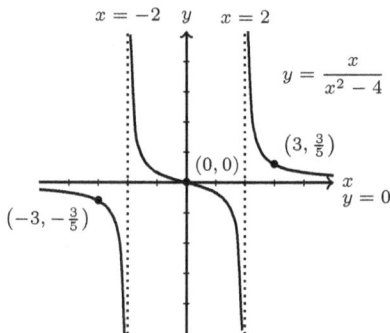

29.

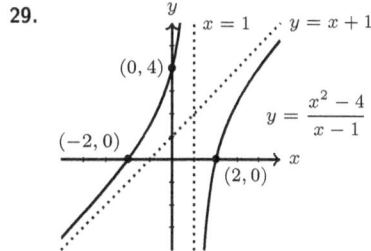

31.

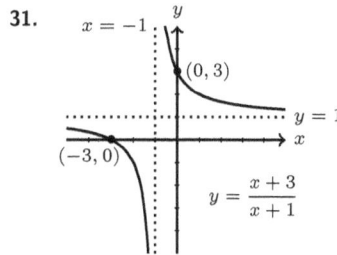

▶ Section 1.2

1. $\dfrac{A}{x - 1} + \dfrac{B}{x + 2}$

3. $\dfrac{A}{x} + \dfrac{B}{x + 2} + \dfrac{C}{(x + 2)^2}$

5. $\dfrac{Ax + B}{x^2 + 1} + \dfrac{Cx + D}{x^2 + 2}$

7. $\dfrac{A}{x} + \dfrac{B}{x^2} + \dfrac{Cx + D}{x^2 + 4} + \dfrac{Ex + F}{(x^2 + 4)^2}$

9. $\dfrac{1}{x} - \dfrac{1}{x + 1}$

11. $\dfrac{1}{4(3x - 4)} - \dfrac{1}{4(3x + 4)}$

13. $\dfrac{4}{x + 3} - \dfrac{2}{x - 3} + \dfrac{1}{(x - 3)^2}$

15. $\dfrac{6}{x^2 + 2} - \dfrac{12}{(x^2 + 2)^2}$

Chapter 2

▶ Section 2.1

1.

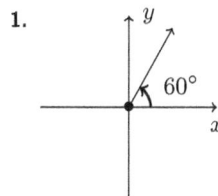

3.

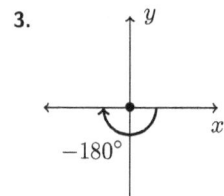

121

5.

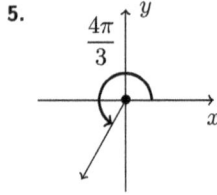

7.

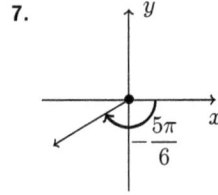

9. 50.31°

11. 45°30′45″

13. $\dfrac{\pi}{9}$

15. $\dfrac{5\pi}{4}$

17. $-\dfrac{\pi}{3}$

19. $-\dfrac{4\pi}{3}$

21. $-\dfrac{11\pi}{6}$

23. 210°

25. 135°

27. −1080°

29. −12°

31. complement: 48°; supplement: 138°

33. complement: 15°; supplement: 105°

35. complement: none because the measure of the angle is greater than 90°; supplement: none because the measure of the angle is greater than 180°

37. 9.42 yd

39. 18.33 cm

41. 0.8 yd^2

43. 66.0 cm^2

45. $\dfrac{\pi}{4}$ in/sec

47. $\dfrac{7}{3}$ rad/sec

▶ **Section 2.2**

1. $\sin\theta = \dfrac{4\sqrt{41}}{41},\quad \cot\theta = \dfrac{5}{4}$

3. $\cos\theta = \dfrac{\sqrt{33}}{7},\quad \tan\theta = \dfrac{4\sqrt{33}}{33}$

5. $\sec\theta = \dfrac{25}{24},\quad \tan\theta = \dfrac{7}{24}$

7. $\sin\theta = \dfrac{5}{13},\quad \cos\theta = \dfrac{12}{13},\quad \tan\theta = \dfrac{5}{12},$
$\csc\theta = \dfrac{13}{5},\quad \sec\theta = \dfrac{13}{12},\quad \cot\theta = \dfrac{12}{5}$

9. $\sin\theta = \dfrac{\sqrt{3}}{2},\quad \cos\theta = \dfrac{1}{2},\quad \tan\theta = \sqrt{3},$
$\csc\theta = \dfrac{2\sqrt{3}}{3},\quad \sec\theta = 2,\quad \cot\theta = \dfrac{\sqrt{3}}{3}$

11. $\sin\theta = \dfrac{2}{3},\quad \cos\theta = \dfrac{\sqrt{5}}{3},\quad \tan\theta = \dfrac{2\sqrt{5}}{5},$
$\csc\theta = \dfrac{3}{2},\quad \sec\theta = \dfrac{3\sqrt{5}}{5},\quad \cot\theta = \dfrac{\sqrt{5}}{2}$

13. $\dfrac{1+\sqrt{3}}{2}$

15. $\dfrac{\sqrt{2}+\sqrt{6}}{4}$

17. $\dfrac{2-\sqrt{3}}{4}$

19. 0.7880

21. 0.4245

23. 1.0223

25. $\dfrac{\sqrt{95}}{12}, \dfrac{12}{7}, \dfrac{\sqrt{95}}{7}, \dfrac{12}{7}$

27. $\dfrac{2}{3}, \dfrac{\sqrt{5}}{3}, \dfrac{2\sqrt{5}}{5}, \dfrac{3}{2}$

29. $\dfrac{2\sqrt{2}}{3}, \dfrac{1}{3}, 2\sqrt{2}, 2\sqrt{2}$

▶ **Section 2.3**

1. $\sin\theta = \dfrac{3\sqrt{10}}{10},\quad \cos\theta = \dfrac{\sqrt{10}}{10},\quad \tan\theta = 3,$
$\csc\theta = \dfrac{\sqrt{10}}{3},\quad \sec\theta = \sqrt{10},\quad \cot\theta = \dfrac{1}{3}$

3. $\sin\theta = -\dfrac{4}{5},\quad \cos\theta = -\dfrac{3}{5},\quad \tan\theta = \dfrac{4}{3},$
$\csc\theta = -\dfrac{5}{4},\quad \sec\theta = -\dfrac{5}{3},\quad \cot\theta = \dfrac{3}{4}$

5. $\sin\theta = \dfrac{1}{2},\quad \cos\theta = -\dfrac{\sqrt{3}}{2},\quad \tan\theta = -\dfrac{\sqrt{3}}{3},$
$\csc\theta = 2,\quad \sec\theta = -\dfrac{2\sqrt{3}}{3},\quad \cot\theta = -\sqrt{3}$

7. III

9. II

11. IV

13. $\sin\theta = -\dfrac{5}{13},\quad \cos\theta = -\dfrac{12}{13},\quad \tan\theta = \dfrac{5}{12},$
$\csc\theta = -\dfrac{13}{5},\quad \sec\theta = -\dfrac{13}{12},\quad \cot\theta = \dfrac{12}{5}$

15. $\sin\theta = \dfrac{3}{5},\quad \cos\theta = -\dfrac{4}{5},\quad \tan\theta = -\dfrac{3}{4},$
$\csc\theta = \dfrac{5}{3},\quad \sec\theta = -\dfrac{5}{4},\quad \cot\theta = -\dfrac{4}{3}$

17. $\sin\theta = \dfrac{7\sqrt{58}}{58},\quad \cos\theta = \dfrac{3\sqrt{58}}{58},\quad \tan\theta = \dfrac{7}{3},$
$\csc\theta = \dfrac{\sqrt{58}}{7},\quad \sec\theta = \dfrac{\sqrt{58}}{3},\quad \cot\theta = \dfrac{3}{7}$

19. 30°

21. 45°

23. $\dfrac{\pi}{3}$

25. $\dfrac{\pi}{4}$

27. $\dfrac{\sqrt{3}}{2}$

29. $-\dfrac{1}{2}$

31. $\dfrac{2\sqrt{3}}{3}$

33. $-\dfrac{1}{2}$

▶ **Section 2.4**

1. $\dfrac{\sqrt{2}}{2}$

3. −1

5. $\sqrt{3}$

7. $\dfrac{\sqrt{3}}{2}$

	Amplitude	Period	Phase Shift	Vertical Shift
9.	2	$2\pi/3$	0	0
11.	3	3π	0	0
13.	3	4π	0	0
15.	2	2π	0	0
17.	1	π	0	−5
19.	1	2π	π	0
21.	3	4π	π	2
23.	3.5	π	$\pi/4$	−1

25. $y = \dfrac{3}{5}\cos\dfrac{4}{3}x$

27. $y = 2\sin(2\pi x + 4\pi)$

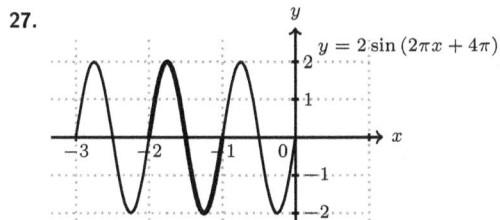

29. $y = \cos(-2x - 2\pi)$

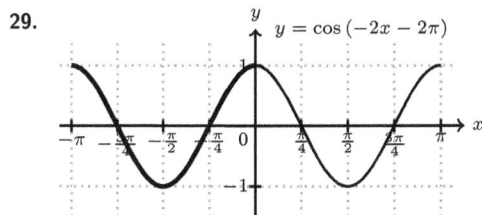

31. $y = \cos(x + \pi) + 1$

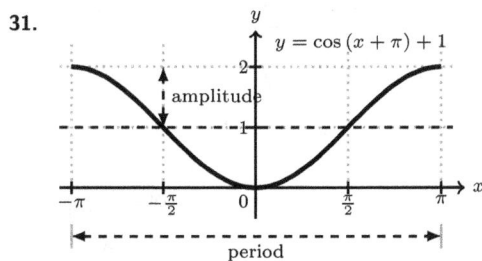

amplitude

period

▶ **Section 2.5**

	Vertical Stretch Factor	Period	Phase Shift	Vertical Shift
1.	1	π	0	2
3.	1	$\pi/2$	$\pi/4$	0
5.	5	$\pi/3$	$-\pi/3$	0
7.	2	2π	0	-3
9.	1	π	$\pi/2$	0
11.	3	$\pi/4$	$-\pi/8$	0

13.

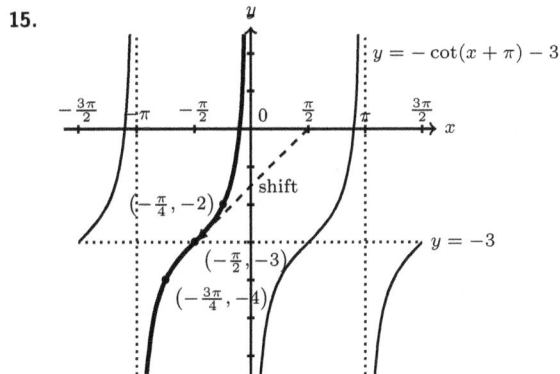

period

$y = 2\cot x - 3$

$\left(\frac{\pi}{4}, -1\right)$

vertical stretch

$\left(\frac{\pi}{2}, -3\right)$

$y = -3$

$\left(\frac{3\pi}{4}, -5\right)$

15.

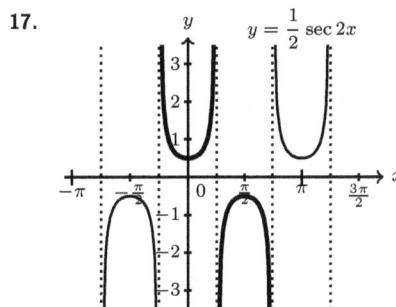

$y = -\cot(x + \pi) - 3$

shift

$\left(-\frac{\pi}{4}, -2\right)$

$\left(-\frac{\pi}{2}, -3\right)$

$\left(-\frac{3\pi}{4}, -4\right)$

$y = -3$

17. $y = \frac{1}{2}\sec 2x$

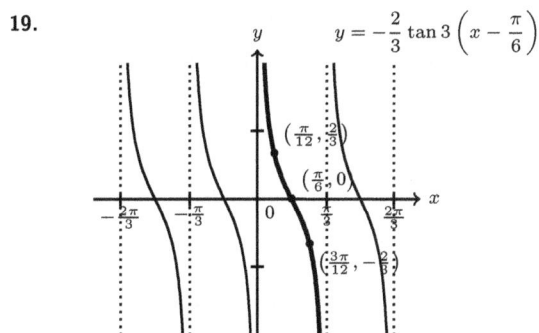

19. $y = -\frac{2}{3}\tan 3\left(x - \frac{\pi}{6}\right)$

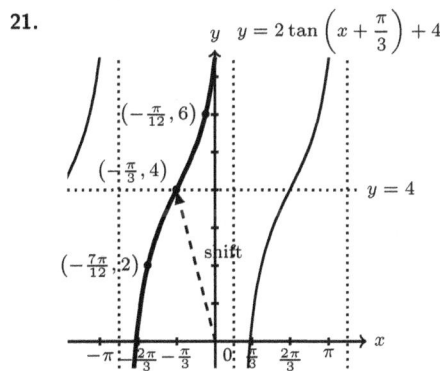

$\left(\frac{\pi}{12}, \frac{2}{3}\right)$

$\left(\frac{\pi}{6}, 0\right)$

$\left(\frac{3\pi}{12}, -\frac{2}{3}\right)$

21. $y = 2\tan\left(x + \frac{\pi}{3}\right) + 4$

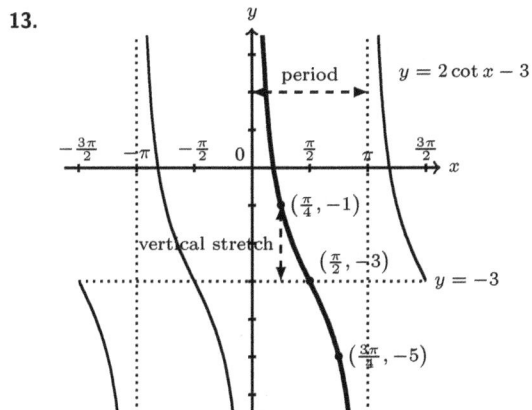

$\left(-\frac{\pi}{12}, 6\right)$

$\left(-\frac{\pi}{3}, 4\right)$

$y = 4$

$\left(-\frac{7\pi}{12}, 2\right)$

shift

▶ **Section 2.6**

1. $\dfrac{\pi}{6}$		**3.** $-\dfrac{\pi}{3}$		**5.** $\dfrac{2\pi}{3}$		**7.** $\dfrac{\pi}{6}$	
9. $\dfrac{\pi}{6}$		**11.** $\dfrac{3}{5}$		**13.** $\dfrac{\pi}{3}$		**15.** $\dfrac{\pi}{3}$	
17. 0		**19.** $\dfrac{12}{13}$		**21.** $\dfrac{\sqrt{2}}{2}$		**23.** $\dfrac{4}{5}$	

25. $-19.1°$ **27.** $-49.6°$

Chapter 3

▶ Section 3.1

1. 1 **3.** $\tan x$ **5.** 1 **7.** $-\tan x$

9. $1 - \tan x$ **11.** $\cos x$ **13.** $\cos x + \sin x$

15. $\text{LHS} = \dfrac{\cos x}{\sin x} + \dfrac{\sin x}{\cos x} = \dfrac{\cos^2 x + \sin^2 x}{\sin x \cos x} = \dfrac{1}{\sin x \cos x}$

$\qquad = \dfrac{1}{\sin x} \cdot \dfrac{1}{\cos x} = \text{RHS}$

17. $\text{LHS} = \dfrac{\cot^2 x}{\csc^2 x} = \dfrac{\cos^2 x}{\sin^2 x} \cdot \dfrac{\sin^2 x}{1} = \text{RHS}$

19. $\text{LHS} = \sin x(1 - \cos^2 x) = \sin x \cdot \sin^2 x = \text{RHS}$

21. $\text{LHS} = \dfrac{\cos^2 x + 1 + 2\sin x + \sin^2 x}{\cos x(1 + \sin x)}$

$\qquad = \dfrac{2 + 2\sin x}{\cos x(1 + \sin x)} = \dfrac{2(1 + \sin x)}{\cos x(1 + \sin x)} = \dfrac{2}{\cos x} = \text{RHS}$

23. $\text{RHS} = \dfrac{1}{\sin^2 x} \cdot \dfrac{\sin^2 x}{\cos^2 x} - 1 = \sec^2 x - 1 = \text{LHS}$

25. $\text{LHS} = \dfrac{1 + \sin x - 1 + \sin x}{1 - \sin^2 x} = \dfrac{2\sin x}{\cos^2 x}$

$\qquad = 2\dfrac{\sin x}{\cos x} \cdot \dfrac{1}{\cos x} = \text{RHS}$

27. $\text{LHS} = (\sin^2 x - \cos^2 x)(\sin^2 x + \cos^2 x) = \sin^2 x - \cos^2 x$

$\qquad = \sin^2 x - (1 - \sin^2 x) = \text{RHS}$

29. $\text{LHS} = \sin^2 x - 2\sin x \cos x + \cos^2 x$

$\qquad\qquad\quad + \sin^2 x + 2\sin x \cos x + \cos^2 x$

$\qquad = 2(\sin^2 x + \cos^2 x) = 2(1) = \text{RHS}$

31. $\text{RHS} = \dfrac{\dfrac{1}{\sin x} \cdot \cos x}{\dfrac{\sin x}{\cos x} + \dfrac{\cos x}{\sin x}} = \dfrac{\dfrac{\cos x}{\sin x}}{\dfrac{\sin^2 x + \cos^2 x}{\cos x \sin x}}$

$\qquad = \dfrac{\cos x}{\sin x} \cdot \dfrac{\cos x \sin x}{1} = \text{LHS}$

33. $\text{LHS} = \sin x + \sin x \csc x = \text{RHS}$

35. $\text{LHS} = \sec^2 x - \tan^2 x = \text{RHS}$

37. $\text{LHS} = \left(\sin x - \dfrac{\sin x}{\cos x}\right)\left(\cos x - \dfrac{\cos x}{\sin x}\right)$

$\qquad = \left(\dfrac{\sin x \cos x - \sin x}{\cos x}\right)\left(\dfrac{\cos x \sin x - \cos x}{\sin x}\right)$

$\qquad = \sin x\left(\dfrac{\cos x - 1}{\cos x}\right)\cos x\left(\dfrac{\sin x - 1}{\sin x}\right) = \text{RHS}$

▶ Section 3.2

1. $\dfrac{\sqrt{6} - \sqrt{2}}{4}$ **3.** $\dfrac{\sqrt{6} + \sqrt{2}}{4}$ **5.** $2 + \sqrt{3}$

7. $\dfrac{\sqrt{2} - \sqrt{6}}{4}$ **9.** $\dfrac{\sqrt{6} + \sqrt{2}}{4}$ **11.** $\dfrac{\sqrt{2} - \sqrt{6}}{4}$

13. $\cos 75°$ **15.** $\cos\dfrac{\pi}{10}$ **17.** $\tan\dfrac{\pi}{5}$

19. $\cot 15°57'$ **21.** $\cos 58°$ **23.** $\tan 80°$

25. $\cos\dfrac{13\pi}{42}$ **27.** $\tan\dfrac{7\pi}{12}$

29. $\text{LHS} = \sin\pi\cos x - \cos\pi\sin x = \text{RHS}$

31. $\text{LHS} = \sin x \cos\dfrac{\pi}{2} + \cos x \sin\dfrac{\pi}{2} = \text{RHS}$

33. $-\cos x$ **35.** $-\sin x$

37. $\dfrac{\sqrt{3}\cos x + \sin x}{\cos x - \sqrt{3}\sin x}$ **39.** $\dfrac{3\sin x + \sqrt{3}\cos x}{3\cos x - \sqrt{3}\sin x}$

41. (a) $\dfrac{16}{65}$ (b) $\dfrac{33}{65}$ (c) $\dfrac{63}{16}$

43. (a) $\dfrac{2 - 2\sqrt{10}}{9}$ (b) $\dfrac{4\sqrt{2} - \sqrt{5}}{9}$ (c) $\dfrac{-8\sqrt{5} - 5\sqrt{2}}{20 - 2\sqrt{10}}$

45. (a) $-\dfrac{36}{85}$ (b) $\dfrac{13}{85}$ (c) $-\dfrac{77}{36}$

▶ Section 3.3

1. $\sin 2\theta = \dfrac{24}{25}, \quad \cos 2\theta = \dfrac{7}{25}, \quad \tan 2\theta = \dfrac{24}{7},$

$\quad \csc 2\theta = \dfrac{25}{24}, \quad \sec 2\theta = \dfrac{25}{7}, \quad \cot 2\theta = \dfrac{7}{24}$

3. $\sin 2\theta = -\dfrac{4\sqrt{5}}{9}, \quad \cos 2\theta = -\dfrac{1}{9}, \quad \tan 2\theta = 4\sqrt{5},$

$\quad \csc 2\theta = -\dfrac{9\sqrt{5}}{20}, \quad \sec 2\theta = -9, \quad \cot 2\theta = \dfrac{\sqrt{5}}{20}$

5. $\sin 2\theta = \dfrac{3}{5}, \quad \cos 2\theta = -\dfrac{4}{5}, \quad \tan 2\theta = -\dfrac{3}{4},$

$\quad \csc 2\theta = \dfrac{5}{3}, \quad \sec 2\theta = -\dfrac{5}{4}, \quad \cot 2\theta = -\dfrac{4}{3}$

7. $\dfrac{1}{2}$ **9.** $\dfrac{\sqrt{3}}{2}$ **11.** $-\dfrac{\sqrt{3}}{2}$ **13.** -1

15. $\text{LHS} = \sin^2 x - 2\sin x \cos x + \cos^2 x = \text{RHS}$

17. $\text{RHS} = \dfrac{1 - \left(1 - 2\sin^2\frac{x}{2}\right)}{2\sin\frac{x}{2}\cos\frac{x}{2}} = \dfrac{\sin\frac{x}{2}}{\cos\frac{x}{2}} = \text{LHS}$

19. $\text{LHS} = \cos(2x + x) = \cos 2x \cos x - \sin 2x \sin x$

$\qquad = (2\cos^2 x - 1)\cos x - 2\sin^2 x \cos x$

$\qquad = 2\cos^3 x - \cos x - 2(1 - \cos^2 x)\cos x = \text{RHS}$

21. $\text{LHS} = \cos[2(2\theta)] = 2\cos^2(2\theta) - 1$

$\qquad = 2(2\cos^2\theta - 1)^2 - 1$

$\qquad = 2(4\cos^4\theta - 4\cos^2\theta + 1) - 1 = \text{RHS}$

23. $\dfrac{\sqrt{2 + \sqrt{3}}}{2}$ **25.** $\dfrac{\sqrt{2 - \sqrt{3}}}{2}$ **27.** $2 + \sqrt{3}$

▶ Section 3.4

1. $\dfrac{1}{2}(\cos x - \cos 5x)$ **3.** $\dfrac{1}{2}\sin 2x$

5. $\dfrac{1}{4} + \dfrac{1}{2}\sin 40°$ **7.** $\dfrac{1}{4}$ **9.** $\dfrac{1}{2} + \dfrac{\sqrt{3}}{4}$

11. $\dfrac{\sqrt{2}}{4}$ **13.** $\dfrac{\sqrt{2}}{4} - \dfrac{1}{4}$ **15.** $\cos 10°$

17. $\sqrt{2}\sin 20°$ **19.** $2\sin 4x \cos x$ **21.** $2\cos 4x \cos x$

23. $\text{LHS} = \dfrac{2\sin 4x \cos(-2x)}{2\cos 4x \cos(-2x)} = \dfrac{\sin 4x}{\cos 4x} = \text{RHS}$

25. $\text{LHS} = \dfrac{-2\sin 6x \sin x}{2\cos 6x \cos x} = -\dfrac{\sin 6x}{\cos 6x} \cdot \dfrac{\sin x}{\cos x} = \text{RHS}$

▶ **Section 3.5**

1. $\dfrac{\pi}{4} + 2k\pi,\ \dfrac{3\pi}{4} + 2k\pi,\ k$: integer

3. $\dfrac{\pi}{6} + k\pi,\ k$: integer

5. $\dfrac{\pi}{6} + 2k\pi,\ \dfrac{5\pi}{6} + 2k\pi,\ k$: integer

7. $\dfrac{\pi}{2} + k\pi,\ k$: integer

9. $\dfrac{\pi}{12},\ \dfrac{19\pi}{12}$ 11. $\dfrac{2\pi}{9},\ \dfrac{14\pi}{9}$

13. $\dfrac{\pi}{4},\ \dfrac{5\pi}{4},\ \dfrac{7\pi}{6},\ \dfrac{11\pi}{6}$ 15. $\dfrac{3\pi}{2}$

17. $\dfrac{\pi}{4},\ \dfrac{3\pi}{4},\ \dfrac{5\pi}{4},\ \dfrac{7\pi}{4}$ 19. $15°,\ 165°,\ 195°,\ 345°$

21. $0°,\ 120°,\ 240°$

Chapter 4

▶ **Section 4.1**

1. $B = 51°,\ a \approx 10.7,\ b \approx 11.8$

3. $B = 23°,\ b \approx 4.0,\ c \approx 8.1$

5. $C = 106°,\ b \approx 15.5,\ c \approx 32.9$

7. $A = 140°,\ b \approx 2.8,\ c \approx 4.6$

9. $C = 59°,\ a \approx 49.1,\ c \approx 43.5$

11. No triangle

13. One triangle, $B \approx 38.1°,\ C \approx 95.9°,\ c \approx 19.4$

15. Two triangles, One $A \approx 88.6°,\ C \approx 55.4°,\ a \approx 17.0$
Two $A \approx 19.4°,\ C \approx 124.6°,\ a \approx 5.7$

17. Two triangles, One $A \approx 34.3,\ B \approx 120.7,\ b \approx 6.1$
Two $A \approx 145.7°,\ B \approx 9.3°,\ b \approx 1.1$

▶ **Section 4.2**

1. $a \approx 22.9,\ B \approx 50.1°,\ C \approx 82.9°$

3. $A \approx 30.8°,\ B \approx 24.1°,\ C \approx 125.1°$

5. $A \approx 27.9°,\ B \approx 125.1°,\ c \approx 3.9$

7. $A \approx 37.9°,\ C \approx 67.1°,\ b \approx 9.4$

9. $B \approx 28.4°,\ C \approx 49.6°,\ a = 20.6$

11. $A \approx 44.4°,\ B \approx 57.1°,\ C \approx 78.5°$

13. $A \approx 93.8°,\ B \approx 29.9°,\ C \approx 56.3°$

15. $A \approx 75.3°,\ B \approx 40.2°,\ C \approx 64.5°$

17. $A \approx 98.2°,\ B \approx 21.8°,\ C \approx 60°$

▶ **Section 4.3**

1. $\mathbf{v} = \langle 1, 3\rangle,\ \mathbf{w} = \langle 0, 2\rangle,\ \mathbf{u} = \langle 0, -2\rangle,\ \mathbf{t} = \langle -4, -2\rangle$

3. 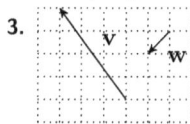   5.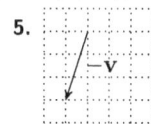

7. 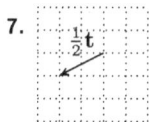   9.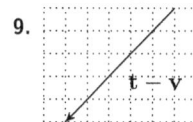

11. $\langle 4, 3\rangle$ 13. $\langle 3, -1\rangle$ 15. $\langle 2, 7\rangle$

17. $\langle -10, -5\rangle$ 19. $\langle 5, -6\rangle$ 21. $\langle 5, 11\rangle$

23. 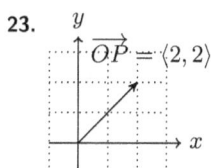   25.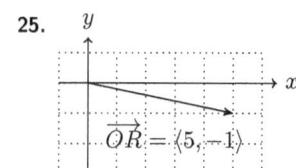

27. $\|\mathbf{v}\| = 2\sqrt{10},\quad \theta = 71.6°$

29. $\|\mathbf{v}\| = 4,\quad \theta = 330°$

31. $\|\mathbf{v}\| = 3,\quad \theta = 180°$

33. $\left\langle -\dfrac{4}{5}, -\dfrac{3}{5}\right\rangle$ 35. $\left\langle -\dfrac{\sqrt{2}}{2}, \dfrac{\sqrt{2}}{2}\right\rangle$

37. $\langle 0, -1\rangle$ 39. $\langle 3\sqrt{3}, 3\rangle$

41. $\left\langle -\dfrac{5\sqrt{2}}{2}, -\dfrac{5\sqrt{2}}{2}\right\rangle$ 43. $\langle 0, 1\rangle$

▶ **Section 4.4**

1. -4 3. 1 5. 1

7. $15\sqrt{2}$ 9. -5 11. $95.2°$

13. $10.3°$ 15. 22.54 17. 4.54

19. Orthogonal   21. Not orthogonal

23. Orthogonal   25. Not orthogonal

27. 76.8 ft-lb   29. $2,969.1$ ft-lb

▶ **Section 4.5**

1. 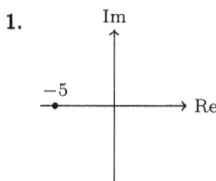   3.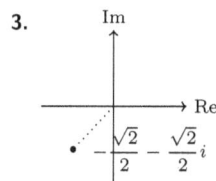

5. 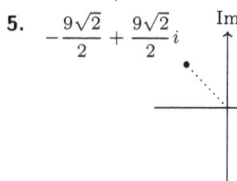   7.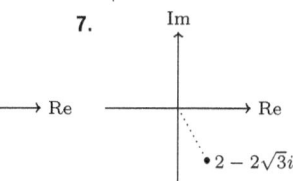

9. $3\sqrt{2}(\cos 315° + i\sin 315°)$

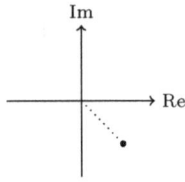

11. $2(\cos 270° + i\sin 270°)$

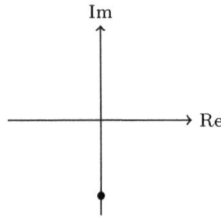

13. $13(\cos 157.4° + i\sin 157.4°)$

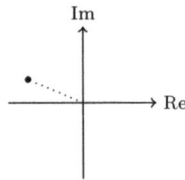

15. $2\sqrt{5}(\cos 243.4° + i\sin 243.4°)$

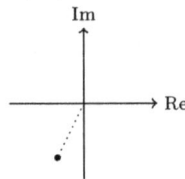

17. $zw = 18(\cos 80° + i\sin 80°)$

$\dfrac{z}{w} = 2(\cos 20° + i\sin 20°)$

19. $zw = 48(\cos 200° + i\sin 200°)$

$\dfrac{z}{w} = \dfrac{3}{4}[\cos(-60°) + i\sin(-60°)]$

21. $zw = 8(\cos 3\pi + i\sin 3\pi)$

$\dfrac{z}{w} = 2(\cos\dfrac{5\pi}{3} + i\sin\dfrac{5\pi}{3})$

23. $25(\cos 80° + i\sin 80°)$

25. $\cos\dfrac{2\pi}{5} + i\sin\dfrac{2\pi}{5}$ **27.** $32 - 32\sqrt{3}i$

29. i **31.** -256

33. $-16 + 16i$ **35.** 1

37. $3(\cos 50° + i\sin 50°)$,
$3(\cos 170° + i\sin 170°)$,
$3(\cos 290° + i\sin 290°)$

39. $\sqrt{3}(\cos\dfrac{\pi}{8} + i\sin\dfrac{\pi}{8}), \sqrt{3}(\cos\dfrac{9\pi}{8} + i\sin\dfrac{9\pi}{8})$

41. $\dfrac{7\sqrt{3}}{2} + \dfrac{7}{2}i, \quad -\dfrac{7\sqrt{3}}{2} - \dfrac{7}{2}i$

43. $\dfrac{\sqrt{2}}{2} + \dfrac{\sqrt{6}}{2}i, \quad -\dfrac{\sqrt{2}}{2} - \dfrac{\sqrt{6}}{2}i$

45. $4i, \quad -2\sqrt{3} - 2i, \quad 2\sqrt{3} - 2i$

Chapter 5

▶ Section 5.1

1. $8, 13, 18, 23, \ldots$ **3.** $0, -2, -2, 0, \ldots$

5. $5, 5, 5, 5, \ldots$ **7.** $1, -\dfrac{1}{4}, \dfrac{1}{9}, -\dfrac{1}{16}, \ldots$

9. $a_n = n$ **11.** $a_n = \dfrac{n+1}{n+2}$

13. $a_n = 2n$ **15.** $a_n = (-1)^n$

17. $a_n = 2^n$ **19.** $1, 4, 7, 10, \ldots$

21. $7, -13, 27, -53, \ldots$ **23.** $1, 1, 2, 6, \ldots$

25. 720 **27.** 15 **29.** $n+1$

31. $(n-2)(n-3)$ **33.** 24

35. $\dfrac{25}{6}$ **37.** $\displaystyle\sum_{n=1}^{98}(n+2)$ or $\displaystyle\sum_{n=3}^{100} n$

39. $\displaystyle\sum_{n=1}^{27} 2n$ **41.** $\displaystyle\sum_{n=1}^{9}(-2)^{n-1}$ or $\displaystyle\sum_{n=0}^{8}(-2)^n$

▶ Section 5.2

1. $a = 1, d = 2$ **3.** not arithmetic

5. $a = 1.2, d = 1$ **7.** $a = \pi, d = \pi$

9. $a_n = -6n + 18, a_{10} = -42$

11. $a_n = n - \dfrac{1}{2}, a_{10} = \dfrac{19}{2}$

13. $a_n = \dfrac{n\pi}{2}, a_{10} = 5\pi$

15. $d = -3, a_n = -3n + 19$

17. $d = 5, a_n = 5n - 25$

19. 30 **21.** 9 **23.** 175

25. $-20,100$ **27.** 2205 **29.** 670

31. 2550 **33.** 10 **35.** $\dfrac{64}{3}$

▶ Section 5.3

1. $a_1 = 2, d = 2$ **3.** $a_1 = 18, d = \dfrac{1}{3}$

5. $a_1 = 1, d = -1$ **7.** not geometric

9. $a_n = 2^n$ **11.** $a_n = (-1)^n$

13. $a_n = (-2)^{n-1}$ **15.** $a_n = 3\left(\dfrac{2}{5}\right)^{n-1}$

17. $a_n = \pi^n$ **19.** 384 **21.** 98

23. $\dfrac{125}{49}$ **25.** 12.5 **27.** 364

29. 8.93 **31.** 66.56 **33.** 124

35. 145.59 **37.** 2935.49 **39.** $\dfrac{9}{2}$

41. 5 **43.** The sum does not exist.

45. 6 **47.** $\dfrac{2}{3}$ **49.** 1000

51. The sum does not exist.

FORMULAS

Sum and Difference Identities

1. $\sin(u+v) = \sin u \cos v + \cos u \sin v$,

   $\sin(u-v) = \sin u \cos v - \cos u \sin v$

2. $\cos(u+v) = \cos u \cos v - \sin u \sin v$,

   $\cos(u-v) = \cos u \cos v + \sin u \sin v$

3. $\tan(u+v) = \dfrac{\tan u + \tan v}{1 - \tan u \tan v}$,

   $\tan(u-v) = \dfrac{\tan u - \tan v}{1 + \tan u \tan v}$

Double Angle Identities

1. $\sin 2x = 2 \sin x \cos x$

2. $\cos 2x = \cos^2 x - \sin^2 x$,

   $\cos 2x = 1 - 2\sin^2 x$,

   $\cos 2x = 2\cos^2 x - 1$

3. $\tan 2x = \dfrac{2 \tan x}{1 - \tan^2 x}$

Power-Reducing and Half-Angle Identities

1. $\sin^2 x = \dfrac{1 - \cos 2x}{2}$, $\quad \sin \dfrac{x}{2} = \pm\sqrt{\dfrac{1 - \cos x}{2}}$

2. $\cos^2 x = \dfrac{1 + \cos 2x}{2}$, $\quad \cos \dfrac{x}{2} = \pm\sqrt{\dfrac{1 + \cos x}{2}}$

3. $\tan^2 x = \dfrac{1 - \cos 2x}{1 + \cos 2x}$,

   $\tan \dfrac{x}{2} = \pm\sqrt{\dfrac{1 - \cos x}{1 + \cos x}} = \dfrac{1 - \cos x}{\sin x} = \dfrac{\sin x}{1 + \cos x}$

Product-to-Sum Identities

1. $\sin x \cos y = \dfrac{1}{2}[\sin(x+y) + \sin(x-y)]$

2. $\cos x \sin y = \dfrac{1}{2}[\sin(x+y) - \sin(x-y)]$

3. $\cos x \cos y = \dfrac{1}{2}[\cos(x+y) + \cos(x-y)]$

4. $\sin x \sin y = \dfrac{1}{2}[\cos(x-y) - \cos(x+y)]$

Sum-to-Product Identities

1. $\sin x + \sin y = 2\sin\left(\dfrac{x+y}{2}\right)\cos\left(\dfrac{x-y}{2}\right)$

2. $\sin x - \sin y = 2\cos\left(\dfrac{x+y}{2}\right)\sin\left(\dfrac{x-y}{2}\right)$

3. $\cos x + \cos y = 2\cos\left(\dfrac{x+y}{2}\right)\cos\left(\dfrac{x-y}{2}\right)$

4. $\cos x - \cos y = -2\sin\left(\dfrac{x+y}{2}\right)\sin\left(\dfrac{x-y}{2}\right)$

The Law of Sines and Cosines

1. $\dfrac{a}{\sin A} = \dfrac{b}{\sin B} = \dfrac{c}{\sin C}$, $\quad \dfrac{\sin A}{a} = \dfrac{\sin B}{b} = \dfrac{\sin C}{c}$

2. $a^2 = b^2 + c^2 - 2bc \cos A$, $\quad \cos A = \dfrac{b^2 + c^2 - a^2}{2bc}$

3. $b^2 = c^2 + a^2 - 2ca \cos B$, $\quad \cos B = \dfrac{c^2 + a^2 - b^2}{2ca}$

4. $c^2 = a^2 + b^2 - 2ab \cos C$, $\quad \cos C = \dfrac{a^2 + b^2 - c^2}{2ab}$

Complex Numbers

1. If $z = a + bi = |z|(\cos\theta + i \sin\theta)$, then

   $|z| = \sqrt{a^2 + b^2}$ and $\tan\theta = \dfrac{b}{a}$.

2. If $z = r(\cos\theta + i \sin\theta)$, then

   $z^n = r^n[\cos(n\theta) + i \sin(n\theta)]$.

3. The nth roots of $r(\cos\theta + i \sin\theta)$ are

   $z_k = \sqrt[n]{r}\left[\cos\left(\dfrac{\theta + 360°k}{n}\right) + i \sin\left(\dfrac{\theta + 360°k}{n}\right)\right]$

   for $k = 0, 1, 2, \ldots, n-1$.

Sequences

1. Arithmetic: $a_n = a_1 + (n-1)d$, $S_n = \dfrac{n}{2}(a_1 + a_n)$,

   $d = \dfrac{a_m - a_n}{m - n}$.

2. Geometric: $a_n = a_1 r^{n-1}$,

   $S_n = \displaystyle\sum_{k=1}^{n} a_1 r^{k-1} = \dfrac{a_1(1 - r^n)}{1 - r}$,

   $S = \displaystyle\sum_{k=1}^{\infty} a_1 r^{k-1} = \dfrac{a_1}{1 - r}$ if $|r| < 1$, and it does not

   exist if $|r| \geq 1$.

The Unit Circle

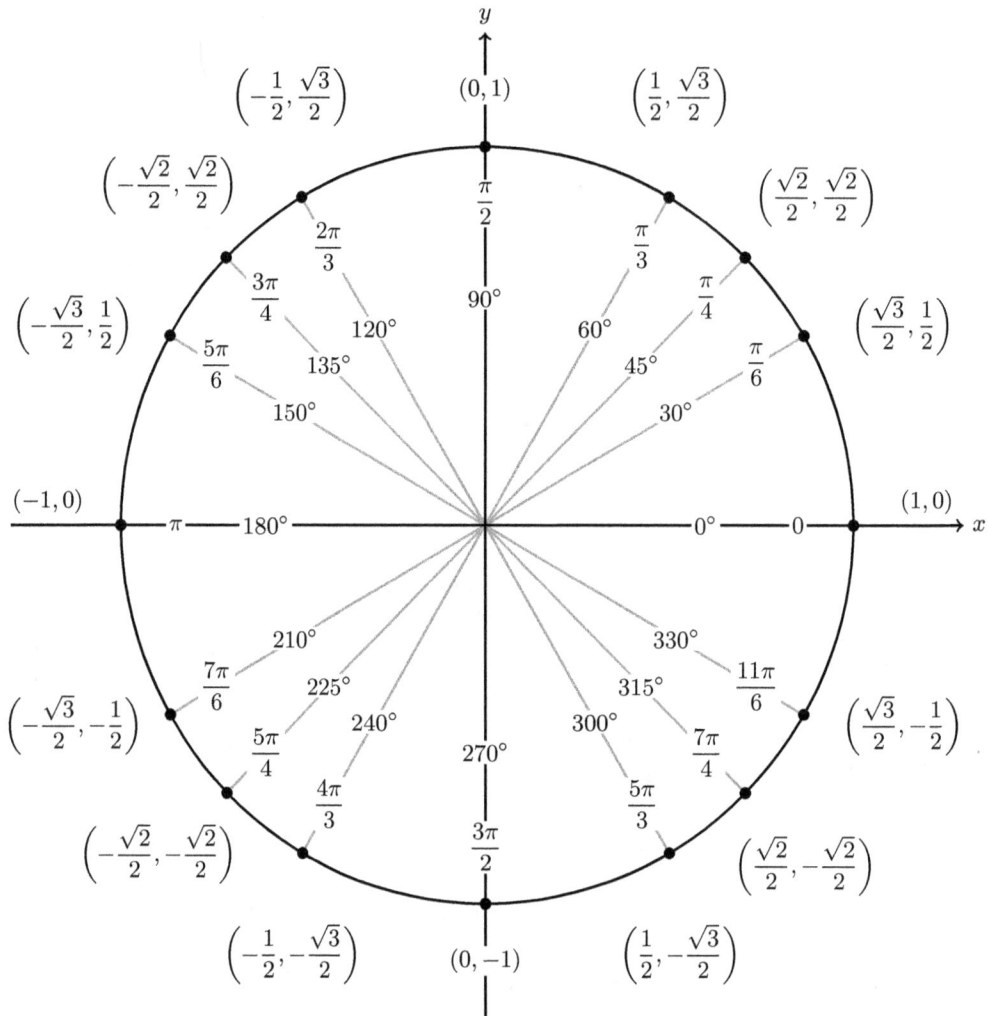

The Unit Circle

$\left(-\frac{1}{2}, \frac{\sqrt{3}}{2}\right)$ $(0,1)$ $\left(\frac{1}{2}, \frac{\sqrt{3}}{2}\right)$

$\left(-\frac{\sqrt{2}}{2}, \frac{\sqrt{2}}{2}\right)$ $\left(\frac{\sqrt{2}}{2}, \frac{\sqrt{2}}{2}\right)$

$\left(-\frac{\sqrt{3}}{2}, \frac{1}{2}\right)$ $\left(\frac{\sqrt{3}}{2}, \frac{1}{2}\right)$

$\frac{\pi}{2}$ $\frac{2\pi}{3}$ $\frac{3\pi}{4}$ $\frac{5\pi}{6}$ $90°$ $120°$ $135°$ $150°$ $\frac{\pi}{3}$ $\frac{\pi}{4}$ $\frac{\pi}{6}$ $60°$ $45°$ $30°$

$(-1,0)$ π $180°$ $0°$ 0 $(1,0)$ x

$210°$ $\frac{7\pi}{6}$ $225°$ $240°$ $\frac{5\pi}{4}$ $\frac{4\pi}{3}$ $270°$ $\frac{3\pi}{2}$ $330°$ $315°$ $300°$ $\frac{11\pi}{6}$ $\frac{7\pi}{4}$ $\frac{5\pi}{3}$

$\left(-\frac{\sqrt{3}}{2}, -\frac{1}{2}\right)$ $\left(\frac{\sqrt{3}}{2}, -\frac{1}{2}\right)$

$\left(-\frac{\sqrt{2}}{2}, -\frac{\sqrt{2}}{2}\right)$ $\left(\frac{\sqrt{2}}{2}, -\frac{\sqrt{2}}{2}\right)$

$\left(-\frac{1}{2}, -\frac{\sqrt{3}}{2}\right)$ $(0,-1)$ $\left(\frac{1}{2}, -\frac{\sqrt{3}}{2}\right)$